FUEL PRICES:
RHYME OR REASON?

FUEL PRICES: RHYME OR REASON?

WILLIAM P. VESTUS
EDITOR

Nova Science Publishers, Inc.
New York

Copyright © 2009 by Nova Science Publishers, Inc.

All rights reserved. No part of this book may be reproduced, stored in a retrieval system or transmitted in any form or by any means: electronic, electrostatic, magnetic, tape, mechanical photocopying, recording or otherwise without the written permission of the Publisher.

For permission to use material from this book please contact us:
Telephone 631-231-7269; Fax 631-231-8175
Web Site: http://www.novapublishers.com

NOTICE TO THE READER

The Publisher has taken reasonable care in the preparation of this book, but makes no expressed or implied warranty of any kind and assumes no responsibility for any errors or omissions. No liability is assumed for incidental or consequential damages in connection with or arising out of information contained in this book. The Publisher shall not be liable for any special, consequential, or exemplary damages resulting, in whole or in part, from the readers' use of, or reliance upon, this material. Any parts of this book based on government reports are so indicated and copyright is claimed for those parts to the extent applicable to compilations of such works.

Independent verification should be sought for any data, advice or recommendations contained in this book. In addition, no responsibility is assumed by the publisher for any injury and/or damage to persons or property arising from any methods, products, instructions, ideas or otherwise contained in this publication.

This publication is designed to provide accurate and authoritative information with regard to the subject matter covered herein. It is sold with the clear understanding that the Publisher is not engaged in rendering legal or any other professional services. If legal or any other expert assistance is required, the services of a competent person should be sought. FROM A DECLARATION OF PARTICIPANTS JOINTLY ADOPTED BY A COMMITTEE OF THE AMERICAN BAR ASSOCIATION AND A COMMITTEE OF PUBLISHERS.

LIBRARY OF CONGRESS CATALOGING-IN-PUBLICATION DATA

Fuel prices : rhyme or reason? / [edited by] William P. Vestus.
  p. cm.
 Includes index.
 ISBN 978-1-60692-842-4 (softcover)
 1. Gasoline--Prices--United States 2. Gasoline industry--United States. 3. Motor fuels--Prices--United States. I. Vestus, William P.
 HD9579.G5U5416 2009
 338.4'3665538270973--dc22
                                                  2009028457

Published by Nova Science Publishers, Inc. ✣ *New York*

CONTENTS

Preface		**vii**
Chapter 1	Gasoline and Oil Prices *Robert Pirog*	1
Chapter 2	The Disparity Between Retail Gasoline and Diesel Fuel Prices *Robert L. Bamberger and Robert Pirog*	17
Chapter 3	Transportation Fuel Taxes: Impacts of a Repeal or Moratorium *Robert Pirog and John W. Fischer*	31
Chapter 4	Motor Fuels: Stakeholder Views on Compensating for the Effects of Gasoline Temperature on Volume at the Pump *United States Government Accountability Office*	41
Chapter 5	Gasoline Prices: Causes of Increases and Congressional Response *Carl E. Behrens and Carol Glover*	65
Chapter Sources		**85**
Index		**87**

PREFACE

The high price of gasoline has been and continues to be a driving factor in consideration of energy policy proposals. Despite passage of the massive Energy Policy Act of 2005 (EPACT 2005, P.L. 109-58), and the Energy Independence and Security Act of 2007 (H.R. 6, P.L. 110-140), numerous other proposed initiatives remain under active consideration in the 110th Congress. Measures proposed include repeal of some tax benefits to domestic oil and gas producers contained in EPACT2005, provisions on price gouging, and reform of oil and gas leasing in the Gulf of Mexico. A large number of factors have combined to put pressure on gasoline prices, including increased world demand for crude oil and limited U.S. refinery capacity to supply gasoline. The war and continued violence in Iraq added uncertainty, and threats of supply disruption have added pressure, particularly to the commodity futures markets. Concern that speculation has added volatility and upward pressure has frequently been cited. This book tries to identify the apparent causes of the wild swings in this most visible of expenses.

Chapter 1 - American gasoline consumers faced rapidly escalating prices during the first half of 2008, though prices began to decline in August. As prices increased to over $4.00 per gallon, consumers faced difficult choices concerning how to allocate limited budgets as the economy slowed. The price increases also adversely affected major industries, including automobile production, transportation, and agriculture. The high gasoline prices also were thought to contribute to the slow-down in economic growth and the potential for general price inflation. The oil industry earned record corporate profits while other sectors of the economy were negatively affected.

Gasoline prices did not increase on their own over this period. The rising price of gasoline was driven by the increasing price of crude oil, the major cost component of gasoline. Crude oil prices, which peaked at just under $150 per barrel in July, and then turned down, rose more quickly than gasoline prices, and the cost share of crude oil per gallon of gasoline rose, putting cost pressure on the refining, distribution, and marketing sectors of the gasoline supply chain.

While the recent increases in the price of crude oil began in late 2007, the price of oil has been increasing, at different rates, since 2004. Many factors have contributed to the price increases over this period. Over the past five years, the mix of factors affecting price at any particular time has varied.

Recently, several factors, including the continuing increase in world oil demand, the effect of speculation on energy futures markets, the transformation of the energy futures market into a pure financial market rather than a commodity market, the declining value of the dollar, foreign governments' fuel subsidization, and limits to the ability of the market to increase supply have been identified as key in explaining oil, and therefore, gasoline price increases.

Policy debates have focused on curbing speculation on oil futures markets by increasing regulatory presence. Over three dozen pieces of legislation have been introduced in the 110th Congress to address speculation-related issues. Other possible policy directions have included declaring a moratorium on collection of the federal excise tax on gasoline, conservation, and the use of oil in the Strategic Petroleum Reserve to augment U.S. supplies of crude oil. Additionally, the possibility of drilling in currently excluded areas on the Atlantic and Pacific outer continental shelf, as well as the Gulf of Mexico, have been considered. The potential opening of the Alaska National Wildlife Refuge for oil exploration and development has also been debated.

The oil market has demonstrated a tendency to be cyclic and sharply volatile. Policy measures that assume long-term stability in the market are unlikely to attain the multiplicity of goals for oil policy the American public desires.

During the summer of 2008, American consumers faced gasoline prices that attained record high levels of over $4.00 per gallon, and oil prices of over $140 per barrel. These high prices have contributed to a downturn in economic growth, and an increase in inflation. They have forced consumers to make difficult choices concerning spending patterns, while their general economic well-being declined. The record prices have raised costs and adversely affected a wide variety of industries, including transportation, automobiles, and agriculture.

Because there does not seem to be one, easily identifiable, factor that has caused these high prices, and because prices rose so quickly from mid-2007 to mid-2008, consensus on how to mitigate the situation through policy has been lacking. Calls for increased exploration and drilling in Alaska and currently restricted offshore areas, energy conservation, increased reliance on alternative energy sources, curbing speculation on oil futures markets, releasing oil from the

Strategic Petroleum Reserve, suspending the federal tax on gasoline, and taxing the profits of oil companies have all been debated.

This chapter examines the extent of price increases in gasoline and oil, focuses on the linkage between the two, and analyzes the causes of the price increases, and the likelihood that they might be reversed through market responses, or policy measures.

Chapter 2 - Over time, gasoline has typically been more expensive than diesel fuel. However, their relative prices have now reversed. In mid-March of 2008, gasoline prices exceeded $3.39/gallon (gal) while diesel fuel prices were above $3.97/gal, a differential of almost $0.60/gal. This has prompted questions of why the historic gap between gasoline and on-highway diesel prices has widened so greatly and over such a relatively brief period of time.

Crude oil, when refined, produces a mix of products. Diesel fuel and home heating oil are derived from the portion of the barrel that produces what are termed "middle distillates." Another part of the barrel furnishes the feedstock for gasoline. Refiners process barrels of crude oil of differing quality, depending on the relative prices for oil of different qualities, and their available technology. Within technology-defined limits refiners can vary the proportions of middle distillate and gasoline production. Because the entire range of petroleum products derive from the same barrel, it is difficult to attribute general refining costs to any single product, making it also difficult to ascertain the relative cost proportions. The exception to this would be when the investment costs of changing product specifications to meet seasonal or environmental requirements can be measured.

A number of specific factors may be identified that have contributed to the shifting relative prices of gasoline and diesel fuel. It is important to recognize that the U.S. market for these fuels is part of a broader world market. World demand patterns are shifting as diesel fuel becomes a primary consumer transportation fuel in Europe and other parts of the world. World price differentials are transmitted to the U.S. market.

Other factors affecting diesel prices include refinery investment costs, as well as investment costs in the product distribution system to accommodate new specifications for diesel fuel that require lower allowable sulfur content; the seasonality of home heating oil demand, a similar product, which transmits the price effects of cold weather from the heating market to the on-highway diesel fuel market; world market effects that might affect the pricing and output mix decisions of refiners; and circumstances affecting the local market at point of purchase.

One other factor should be noted. The primary demand sectors for gasoline and diesel fuel are different in the United States. Gasoline is a mass consumer good and home heating oil an important regional and seasonal residential product, while diesel fuel is used in a wide variety of commercial and industrial applications. Diesel fuel is often part of the cost of delivering goods and providing services. As a consequence, demand for diesel fuel may be less elastic, and therefore, likelier to be passed on to consumers.

By mid-March of 2008, gasoline prices exceeded $3.39/gallon (gal) while diesel fuel prices were $3.97/gal, a differential of almost $0.60/gal. In mid-March of 2007, the relationship between the two fuels was the reverse: gasoline prices were higher than diesel prices. At that time, diesel prices were roughly $2.68/gal, while the average price of gasoline for all grades was $2.76 — more than $0.08 higher than the average price of on-highway diesel. Additionally, where gasoline prices in mid- March 2008 are roughly $0.63/gal higher than year-ago averages, diesel fuel prices have risen over $1 .29/gal over the same period. Over $0.60/gal of this increase has occurred since the beginning of 2008. This has prompted questions of why the historic gap between gasoline and on-highway diesel prices has widened so greatly and over such a relatively brief period of time. Because diesel fuel costs affect the cost of shipping by truck, price increases affect the delivered cost of most consumer goods purchased in the United States, contributing to the over-all level of price inflation.

This chapter provides background and identifies some of the likely factors and forces in world markets that may have contributed to the evolution of the relative prices of gasoline and diesel fuel over the past several years. Among these are strong international demand for diesel fuel; product mix decisions by refiners, and refinery investment to meet more stringent limits on the sulfur content of diesel fuel; the similarities between diesel fuel and home heating oil; and the effect on retail prices from local market conditions.

Chapter 3 - Legislation that would repeal or otherwise provide for a summer-long moratorium of federal transportation fuel taxes has been introduced in the 1 10th Congress. Simultaneously, Senators McCain and Clinton are proposing a summer fuel tax collection moratorium as part of their Presidential campaigns. Fuel prices have risen rapidly in 2008 for a variety of reasons. Those seeking to alter federal fuel tax collection are doing so in the belief that a reduction in fuel taxes would give Americans a modest level of economic relief from high pump prices. Current market conditions and the marginal amount of tax relief incorporated in most proposals, however, raise uncertainty as to whether prices to individuals and businesses would fall and whether any price decline would be meaningful to consumers in economic terms. Also of concern is the possible impact of any repeal or moratorium on the overall federal budget deficit.

A reduction in transportation fuel taxes would result in a decrease in spending for Highway Trust Fund-supported federal programs, unless Congress designated alternate sources of funding for these programs. As a result of the structure of the federal programs, the effects of a fuel tax repeal on federal transportation programs would not necessarily be immediate, but depending on the length and scope of the repeal or suspension, they could be substantial.

Legislation that would repeal or otherwise provide for a summer-long moratorium of federal transportation fuel taxes has been introduced in the 110th Congress. Simultaneously, Senators McCain and Clinton are proposing a summer fuel tax collection moratorium as part of their Presidential campaigns. Fuel prices have risen rapidly in 2008 for a variety of reasons. Those seeking to alter federal fuel tax collection are doing so in the belief that a reduction in fuel taxes would give Americans a modest level of economic relief from high pump prices. There is, however, significant opposition to this proposal by those who believe that a fuel tax "holiday" would provide minimal relief to individuals, while potentially adding to the overall federal deficit.

Chapter 4 - The volume, but not the energy content, of hydrocarbon fuels, such as gasoline and diesel, varies in response to changes in temperature. Thus, because of expansion, the energy content per gallon of 90 degree fuel is less than that of 60 degree fuel. States and localities adopt and enforce weights and measures regulations, often using the model regulatory standards published by the National Institute of Standards and Technology (NIST). Although technology now exists to compensate for the effects of temperature on gas volume, the costs of doing so at the retail level have become the subject of much debate among weights and measures officials, consumer groups, and representatives of the petroleum and fuel marketing industries.

GAO was asked to provide information on (1) the views of U.S. stakeholders on the costs to implement automatic temperature compensation, (2) the views of U.S. stakeholders on who would bear these costs, and (3) the reasons some state and national governments have adopted or rejected automatic temperature compensation. To do this work, GAO reviewed NIST and other documents and congressional testimony; interviewed stakeholders from 3 federal agencies, 17 states, and 15 groups representing a variety of interests, including consumers, truck drivers, and the oil and gas industry; and interviewed officials in 5 other nations.

Chapter 5 - The high price of gasoline has been and continues to be a driving factor in consideration of energy policy proposals. Despite passage of the massive Energy Policy Act of 2005 (EPACT 2005, P.L. 109-58), and the Energy Independence and Security Act of 2007 (H.R. 6, P.L. 110-140), numerous other proposed initiatives remain under active consideration in the 1 10th Congress. Measures proposed include repeal of some tax benefits to domestic oil and gas producers contained in EPACT2005, provisions on price gouging, and reform of oil and gas leasing in the Gulf of Mexico.

A large number of factors have combined to put pressure on gasoline prices, including increased world demand for crude oil and limited U.S. refinery capacity to supply gasoline. The war and continued violence in Iraq added uncertainty, and threats of supply disruption have added pressure, particularly to the commodity futures markets. Concern that speculation has added volatility and upward pressure has frequently been cited. In recent months, a decline in the value of the dollar compared to other currencies has increased the dollar price of oil on futures markets.

The gasoline price surge has stimulated much legislative activity, but until recently there has not been the sense of the extreme urgency of previous energy crises. In part, this may be due to the fact that there has been no physical shortage of gasoline or lines at the pump, as there were after the Arab oil embargo in 1973 and the Iranian revolution in 1979. At that time there was expectation that prices were destined to grow ever higher, and many believed that the world's supply of oil was running out. Such views have been less prevalent during the current run-up. But the continued and unrelenting increase in crude oil prices to record levels, even discounting inflation, is leading many to suggest that changing world market conditions may have led to permanent, or at least chronic, shortages of petroleum production capacity. Others continue to expect that growth in demand will moderate, and production will increase to meet demand, as it did following the shortages of the 1970s.

The continuing high prices have led to a further search for legislative remedies. This chapter, after analyzing factors that have contributed to high gasoline prices, describes the major legislative initiatives and discusses the issues involved.

In: Fuel Prices: Rhyme or Reason?
Editor: William P. Vestus

ISBN: 978-1-60692-842-4
© 2009 Nova Science Publishers, Inc.

Chapter 1

GASOLINE AND OIL PRICES

Robert Pirog

SUMMARY

American gasoline consumers faced rapidly escalating prices during the first half of 2008, though prices began to decline in August. As prices increased to over $4.00 per gallon, consumers faced difficult choices concerning how to allocate limited budgets as the economy slowed. The price increases also adversely affected major industries, including automobile production, transportation, and agriculture. The high gasoline prices also were thought to contribute to the slow-down in economic growth and the potential for general price inflation. The oil industry earned record corporate profits while other sectors of the economy were negatively affected.

Gasoline prices did not increase on their own over this period. The rising price of gasoline was driven by the increasing price of crude oil, the major cost component of gasoline. Crude oil prices, which peaked at just under $150 per barrel in July, and then turned down, rose more quickly than gasoline prices, and the cost share of crude oil per gallon of gasoline rose, putting cost pressure on the refining, distribution, and marketing sectors of the gasoline supply chain.

While the recent increases in the price of crude oil began in late 2007, the price of oil has been increasing, at different rates, since 2004. Many factors have contributed to the price increases over this period. Over the past five years, the mix of factors affecting price at any particular time has varied.

Recently, several factors, including the continuing increase in world oil demand, the effect of speculation on energy futures markets, the transformation of the energy futures market into a pure financial market rather than a commodity market, the declining value of the dollar, foreign governments' fuel subsidization, and limits to the ability of the market to increase supply have been identified as key in explaining oil, and therefore, gasoline price increases.

Policy debates have focused on curbing speculation on oil futures markets by increasing regulatory presence. Over three dozen pieces of legislation have been introduced in the 110th Congress to address speculation-related issues. Other possible policy directions have included declaring a moratorium on collection of the federal excise tax on gasoline, conservation, and the use of oil in the Strategic Petroleum Reserve to augment U.S. supplies of crude oil. Additionally, the possibility of drilling in currently excluded areas on the Atlantic and Pacific outer continental shelf, as well as the Gulf of Mexico, have been considered. The potential opening of the Alaska National Wildlife Refuge for oil exploration and development has also been debated.

The oil market has demonstrated a tendency to be cyclic and sharply volatile. Policy measures that assume long-term stability in the market are unlikely to attain the multiplicity of goals for oil policy the American public desires.

During the summer of 2008, American consumers faced gasoline prices that attained record high levels of over $4.00 per gallon, and oil prices of over $140 per barrel.1 These high prices have contributed to a downturn in economic growth, and an increase in inflation. They have forced consumers to make difficult choices concerning spending patterns, while their general economic well-being declined. The record prices have raised costs and adversely affected a wide variety of industries, including transportation, automobiles, and agriculture.

Because there does not seem to be one, easily identifiable, factor that has caused these high prices, and because prices rose so quickly from mid-2007 to mid-2008, consensus on how to mitigate the situation through policy has been lacking. Calls for increased exploration and drilling in Alaska and currently restricted offshore areas, energy conservation, increased reliance on alternative energy sources, curbing speculation on oil futures markets, releasing oil from the Strategic Petroleum Reserve, suspending the federal tax on gasoline, and taxing the profits of oil companies have all been debated.

This chapter examines the extent of price increases in gasoline and oil, focuses on the linkage between the two, and analyzes the causes of the price increases, and the likelihood that they might be reversed through market responses, or policy measures.

GASOLINE AND OIL PRICE INCREASES

The Energy Information Administration (EIA) reported that retail gasoline prices increased by 33% from January through July 2008, or, by over $1 per gallon.[2] Although the price was higher in each month in 2008, on a month-by-month basis, compared to 2007, the greatest differential was observed in July, when prices were about $1.10 per gallon higher than in the previous year.

These price increases led consumers to respond by using less gasoline and driving fewer miles. By mid-July 2008, gasoline demand had declined by about 340 thousand barrels per day (b/d), or 3.6%, compared to a similar period in 2007.[3] Miles traveled declined by 3.7% in May 2008 compared to May 2007.[4] These responses are important because they represent a potential market adjustment that could result in the moderating of the price increases of the first half of 2008. If consumption did not decline in the face of sharp price increases, there would be little incentive for producers not to continue raising prices.

Table 1. U.S. Monthly Gasoline Prices (cents per gallon).

	2007	2008
January	224.0	304.3
February	227.8	302.8
March	256.3	324.4
April	284.5	345.9
May	314.6	376.6
June	305.6	405.4
July	296.5	406.2
August	278.6	
September	280.3	
October	280.3	
November	308.0	
December	301.8	

Source: Energy Information Administration, available at [http://www.eia.doe.gov].

During the price increases of the first half of 2008, the cost composition of gasoline changed. **Table 2** shows that not only did crude oil cost more per gallon of gasoline, but its share of the total price increased. Crude oil cost $1.54 per gallon of gasoline in June 2007, and $3.00 per gallon of gasoline in June 2008.

Table 2. Cost Composition of Gasoline, June 2007-2008 (percent).

	2007	2008
Crude oil	50.5	74.0
Refining	22.7	9.0
Distribution/Marketing	13.7	5.0
Taxes	13.0	10.0

Source: Energy Information Administration, *Gasoline and Diesel Fuel Update*, July 28, 2008.

Table 3. Monthly West Texas Intermediate Oil Spot Price (dollars per barrel).

	2007	2008
January	$54.51	92.97
February	59.28	95.39
March	60.44	105.45
April	63.98	112.58
May	63.45	125.40
June	67.49	133.88
July	74.12	133.37
August	72.36	
September	79.91	
October	85.80	
November	94.77	
December	91.69	

Source: Energy Information Administration, available at [http://www.eia.doe.gov].

Refining, as a share of the cost of a gallon of gasoline declined by over 50%, and fell from $0.69 per gallon in 2007, to $0.36 per gallon in 2008. This nearly 48% fall in cash flow to refiners has reversed the economic performance of refiners in recent years, and puts into even greater question the possibility of expanding the U.S. refining capacity base. Distribution and marketing's share of the price of a gallon of gasoline also declined from June 2007 to 2008 putting pressure on gas station owners, as they saw their net return per gallon sold fall. Taxes were less affected by the growing role of crude oil in the cost structure of gasoline because, while the federal excise tax on gasoline is a fixed $0.184 per gallon, and as a result will decline in percentage terms as the price of gasoline rises, some state and local taxes on gasoline are percent of value, or *ad valorem,* taxes and will retain a constant share of cost as the price of gasoline rises.

The data presented in **Table 2** suggests that the major reason the price of gasoline has risen in 2008 is because the price of crude oil has risen. The rise in the price of crude oil has not only affected consumers, it has affected virtually all the parts of the gasoline supply chain.

Table 2 and **Table 3** show that while the price of crude oil rose by about 24% in the second half of 2007, the price of gasoline rose by only about 2% over the same time period. While crude oil prices increased by another 44% in the first half of 2008, gasoline prices increased by 34% over the same period. Anticipated and actual cuts in demand for gasoline may have caused firms to resist passing the full cost increases of crude oil on to consumers in 2007, increasing the rate at which gasoline prices rose in early 2008.

FACTORS AFFECTING OIL PRICES

The price of oil is set on a world market, over which no firm has direct control. However, the market differs in significant ways from the economic conception of a free, competitive, market. Five countries-the United States, China, Japan, India, and South Korea-consumed over 45% of the world demand of 85.2 million barrels per day in 2007, while producing only 14% of world production. As a result, these countries, as well as many other net consuming nations, depend on world trade in crude oil. Since production of crude oil depends on a geological formation to yield the oil, only certain places in the world can produce oil at high levels, in excess of their own consumption. These areas are concentrated in the Persian Gulf, where over 60% of known proved reserves are located. Reserves can be augmented, and production can be increased, only after the expenditure of billions of dollars and years of development after a discovery. The Persian Gulf oil producers and others attempt to exercise control of the price of oil through the Organization of the Petroleum Exporting Countries (OPEC).

Because of the characteristics of consumers and producers in the market, both the demand and the supply sides of the market are very inelastic, or price insensitive, in the short and intermediate term. Since neither consumers nor producers can easily make quantity adjustments in response to changing prices, the market price may not be singular. There may be a range of prices that are consistent with the market avoiding physical shortages or surpluses. In this type of market environment, high levels of price volatility are likely.

WORLD DEMAND

A frequently cited reason for high crude oil prices is the growth of world demand, driven by China and India.

Table 4 shows that the total growth in demand over the past five years of 7.5% is higher than the total growth of 5.3% over the previous five years. Over the past five years, from 2003 to 2007, consumption growth in China totaled 35%, higher than the world growth rate. However, the growth of Chinese demand cannot be considered in isolation. Since the crude oil market is world-wide, the high level of Chinese growth should be considered only in the context of total world demand growth.[5]

Although 1997 and 2004 stand out as years in which annual growth exceeded the ten year average of 1.3%, the data in **Table 4** might not appear to suggest that demand growth could be sufficient to push crude oil prices to the record levels observed in July 2008. In the case of the oil market, it may be that percentage changes taken alone are misleading.

Table 4. World Oil Demand, 1997-2007 (thousand barrels per day).

Year	Demand	% of Growth	Year	Demand	% of Growth
1997	73,598	5.8	2003	79,296	1.9
1998	73,939	0.05	2004	82,111	3.5
1999	75,573	2.2	2005	83,317	1.5
2000	76,340	1.0	2006	84,230	1.1
2001	76,904	0.07	2007	85,220	1.2
2002	77,829	1.2			

Source: BP Statistical Review of World Energy 2008.
Note: Growth is the percentage change in a year over the previous year.

Because of the nature of oil production, which is characterized by time lags, an inability to easily expand output from existing fields, and low incentives to keep production as excess capacity, the actual volume of demand increases, irrespective of the percentage value, is a key factor. This differentiates the oil industry from other manufacturing and service industries where marginal percentage increments in output can be met by using the existing capital stock and labor force more intensively.

The demand increase of 3.5%, or over 2.8 million b/d, in 2004, reduced spare capacity in the world, and created a tight balance between demand and supply. When growth fell to 1.5% in 2005, that, nonetheless, represented a further expansion of required world production by 1.2 million b/d. Because world production could not expand quickly enough to meet this new demand, further reductions in excess capacity occurred and the tight market conditions continued. When an increase in demand causes total demand to exceed current production, it must be met by reducing excess capacity. When excess capacity in the industry falls, markets anticipate that the reduced excess capacity is less able to accommodate future possible supply disruptions and, as a result, the price increases.

The solution to rising prices, assuming that world demand growth cannot be controlled, is expansion of output. However, the largest oil reserves are believed to be already discovered and in production. Future fields might well be smaller and more costly. Data shows that in 2007, the world level of proven reserves declined from 1.239 trillion barrels to 1.237 trillion barrels, suggesting that new discoveries did not offset 2007 production levels.[6]

FINANCIAL SPECULATION

One of the most debated factors in rising oil prices has been the role of speculators, particularly those investing through commodity index funds. Speculators have always been part of energy futures markets, but it generally has been as the other side of hedging transactions by physical traders.[7] Their role has been to accept price risk when physical traders sought to transfer it, and lock-in prices. Recently, however, commodities' futures markets have become part of financial investor portfolio strategies. As a result, the market has seen an inflow of new participants that have no interest in the physical commodity, beyond the possibility of profiting from its price variations.

The oil futures market is actually composed of three different markets; regulated futures exchanges like the New York Mercantile Exchange (NYMEX), electronic trading facilities like the Intercontinental Exchange (ICE), and the Over the Counter (OTC), or swap market. Each market offers a different contract, traded under different rules, and with different degrees of regulatory control.[8] The focus for oil price effects has recently been on the NYMEX where a standardized futures contract, based on West Texas Intermediate (WTI) crude oil, deliverable at Cushing, Oklahoma, is traded under the regulation of the Commodities Futures Trading Commission (CFTC). This market is financial in nature, in the sense that little or no oil actually is traded, or changes hands. Almost all contract gains and losses are settled in cash, and positions, either long (as a holder of the right to buy oil) or short (as a holder of the right to sell oil) can be closed out, or offset by purchasing a contract that offsets the original position.

In principle, there is little reason why any transaction on the oil futures markets should affect the current price of oil. What is being traded in these markets is a contract that obligates an investor to buy oil one month, or more, in the future at a known, set, price. If more investors desire to be in that position, the demand for the contract goes up, taking with it the contract price at which that oil might be traded. Again, in principle, this transaction does not affect any demand or supply fundamental in the physical oil market, and should not necessarily affect the current price of oil. In general, it is more likely that shifts in the underlying fundamentals of oil demand and supply could affect the expectations of future prices held by investors, and alter their behavior causing them to buy or sell, leading to price variations that reflect the evolving market forces in the physical market.

In the real market for oil, it is possible that purely financially based expectations, realized through changing positions in the financial markets, might affect the real price of oil. The spot market, a market for the delivery of real crude oil, bases its price on the futures market, and EIA data show that the two prices vary only by a few cents as observed in **Table 5**.

Table 5. Futures and Spot Prices for WTI Crude Oil, 2007-2008 (dollars per barrel).

	Spot price 2007	Future price 2007	Spot price 2008	Future price 2008
January	$54.51	$54.35	$92.97	$92.93
February	59.28	59.39	95.39	95.35
March	60.44	60.74	105.45	105.45
April	63.98	64.04	112.58	112.46
May	63.45	63.53	125.40	125.46
June	67.49	67.53	133.88	134.02
July	74.12	74.15		
August	72.36	72.36		
September	79.91	79.63		
October	85.80	85.55		
November	94.77	94.63		
December	91.69	91.74		

Source: Energy Information Administration, available at [http://www.eia.doe.gov].

The reason for this close correlation is that the spot price is in itself a forward price. Oil contracted today on the spot market must generally be delivered within twenty-one days, with notice given to the buyer prior to delivery. The near month futures contract generally covers thirty days or less into the future, hence there issubstantial overlap between futures market contract delivery and spot market delivery. As a result, the NYMEX near month futures price is the basis for the spot market price, adjusted for relatively minor differences in delivery time and other factors. This linkage between the spot and futures market price is the connection between the financial oil market and the real oil market. It allows investment decisions by financial institutions and investment funds to be transferred quickly and directly to gasoline consumers.

It is possible to take almost the reverse position on the spot versus future price correlation. In this view, the futures price must always adjust to the spot price. At the expiration date of the futures contract, the spot price must equal the futures price on the expiring contract because the holder of futures contracts can always choose to receive oil rather than a cash settlement if desired. Futures prices are seen as driven by expectations of the future spot price. In this market conception, the primary control mechanism is that essentially all of the futures market participants are real commodity hedges.[9]

Several factors are important in balancing these arguments. The approach that claims futures markets control the spot price are based on investor behavior that is at odds with that of the perfectly rational financial investor of economic theory. For this approach to have validity, two tests should be met. First, there should be a class of participants in the market that are not hedgers in the sense of having a need to lock in the price of oil for commercial use, for example, purely financial investors. Second, there should be some explanation, and evidence, that the non-commercial class of market participants were likely making investment decisions driven by some other rationale beyond that of rational economic, financial analysis.

No definite, quantitative evidence is available for either condition; however, testimony given at congressional hearings indicates that new participants may be entering the oil futures markets in the form of investment by commodity index funds.[10]

The data in **Table 6** show that over a five year period, holdings of crude oil and gasoline in the form of futures contracts, by investors in commodity index funds increased by over 500% in the case of WTI crude oil, and over 200% in the case of gasoline. Investors in these funds include pension funds, university endowments, private investors, hedge funds and sovereign wealth funds. Commodity index funds pool investors' funds and purchase futures contracts in a wide variety of commodities markets, including agriculture, livestock, energy, base, and precious metals. In each of these markets, the funds have expanded their holdings since 2003 in amounts comparable to those observed in **Table 6**.[11]

Table 6. Commodity Purchases by Index Speculators (millions of barrels).

	Holdings 1/1/03	Holdings 3/12/08	Net Purchases
WTI Crude Oil	99.88	638.38	538.99
Brent Crude Oil	47.07	191.59	144.52
Gasoline	2,624.24	8,549.15	5,924.90

Source: Michael W. Masters, Testimony before the Committee on Homeland Security and Governmental Affairs, United States Senate, May 20, 2008.

The purpose of commodity index fund investment is not to gain title to commodities; the funds settle their contract positions in cash. The goal is the benefit of diversification as well as the potential for high rates of return. Commodity prices are thought to vary inversely with financial investments in traditional corporate shares and bonds. For example, if the Dow Jones Industrial Average of corporate shares declines on a particular day, investors might expect to observe an increase in the index value of commodity futures. This inverse performance of financial and commodity markets reduces the over-all risk in investors' portfolios, and presents investors a more favorable risk/return profile.

Since futures contracts in energy markets have a finite time horizon, when the contracts expire, new contracts must be purchased to keep the target balance between stocks, bonds, and commodities in place. As a result, once commodity index fund managers determine that commodity investments are a desirable part of their portfolios, this new source of demand is, in effect, permanent. Any new source of demand, driven not by the fundamentals, or expectations, of oil demand and supply, but by portfolio decisions is likely to raise the price of an oil contract on the futures market, which is transmitted directly to consumers through the linkage between the futures and spot market prices.

For commodity index investments to be an important factor in the increasing price of oil over time, not only must there be an initial expansion of demand, but growing demand in each subsequent time period. This type of demand growth might result from favorable expectations concerning the fundamentals of the oil market, or it could result from a herd mentality among portfolio managers, which could contribute to a financial bubble due to speculators entering the commodity futures market and driving up prices of contracts. If the commodity futures market has a bias toward this type of upward price movement, and it is supported by strong fundamentals in the real commodity markets, growing demand and/or tight supplies, it is likely that the result will be increasing prices, as observed in the price of oil since 2004.

VALUE OF THE DOLLAR

Many analysts believe the price of oil varies inversely with the value of the dollar against other major currencies, notably the euro. Oil is priced in dollars on the international market. If the dollar falls in value against another currency, it takes fewer units of that currency to purchase the requisite number of dollars needed to buy a barrel of oil. As a result, oil becomes cheaper in the other currency, say the euro, than in dollars. The price of oil measured in dollars then rises. Each dollar purchases a smaller fraction of a barrel of oil as a result. Holding oil, rather than dollars, becomes a way to protect against a declining dollar.

Over the period of rising oil prices, 2004 to mid-2008, the euro/dollar exchange rate has been variable. From January 2, 2004 to January 3, 2005, the euro increased in value relative to the dollar, by 7%. From January 3, 2005 to January 3, 2006, the euro declined in value relative to the dollar by 11%. From January 3, 2006, to January 2, 2007 the euro increased in value relative to the dollar by 10.8%. From January 2, 2007 to January 2, 2008 the euro increased in value relative to the dollar by 10.9%.[12]

The increasing value of the euro in 2007 may have been a factor in the increasing price of oil over the same period.

NATIONAL SUBSIDIES

In many areas of the world, gasoline consumers have been protected from the effects of higher world oil prices by their governments through fuel subsidies. For example, it was reported that in July 2008 the price for a gallon of gasoline in Venezuela was $0.12, $0.40 in Iran, $0.45 in Saudi Arabia, and $0.89 in Egypt.[13] High population countries like China, Indonesia, and Russia also subsidize gasoline.

The effect of fuel subsidies is to reduce the ability of the natural market forces of demand to lower price. Consumers make decisions based on artificially set prices. Consumption is higher than it would be if they faced the market price. Producers see less reason to lower prices because demand continues to be strong. The demand effect also affects consumers in nations that do not subsidize. Although consumers in the non-subsidy countries may cut back on consumption when faced with higher prices, their decisions may be negated by demand growth in subsidized markets. In this way, governments that subsidize fuel costs in their domestic markets are adopting a strategy which supports oil producers and high prices.

Nations that subsidize fuel costs do pay a price, whether they are oil exporters or not. The subsidy is generally a large drain on the national treasury, causes inefficient allocation of resources, and anomalies in trade. For example, Iran exports over 2 million barrels per day of crude oil, but imports gasoline because of domestic demand for gasoline that exceeds Iran's refining capacity.

OIL SUPPLY

Oil supply is relatively insensitive to price changes in the short run. More oil production can enter the market in times of high prices only if excess capacity exists at producing fields, given the level of demand.[14] Similarly, production only minimally declines when prices fall, because it is uneconomic to cut production rates at producing fields, due to the wide spread between the cost of extraction and typical levels of price observed over the past five years, as well as high fixed costs. Excess capacity is thought to exist only in the Organization of the Petroleum Exporting Countries (OPEC).

Excess capacity hit a recent peak of about 7 million barrels per day in 2002, and has been lower than peak, but increasing, since 2004. **Table 7** shows the pivotal role played by Saudi Arabia in the world oil market. Saudi excess capacity accounted for 82% of OPEC spare capacity in 2005, and 61% in 2007. Nigerian oil production is subject to disruption due to political upheaval, and Iran and the oil importing nations are at odds over Iran's nuclear ambitions. The most direct way to augment spare capacity in the short-run is through demand reduction, generally as a result of higher prices and slowing economic growth. Because excess capacity has increased over the 2007-2008 period, it is unlikely that excess capacity constraints were the major factor in the recent price increases.

In the longer term, if world demand for oil continues to grow as more nations develop and gain income and wealth, excess capacity is likely to be tight, unless high levels of investment in productive capacity take place. Investment in productive capacity is limited by "resource nationalism." Resource nationalism refers to the practices of countries with known, or suspected, oil reserves that limit access to those potential supplies. If national oil companies, either due to lack of funding, or expertise, or political direction, fail to develop oil resources the world oil market is likely to have difficulty keeping up with world demand, perpetuating high prices.

Table 7. OPEC Excess Crude Oil Capacity (thousands of barrels per day).

	2005	2006	2007	2008[a]	2010[a]
Saudi Arabia	1,473	2,032	2,673	3,456	3,218
Angola	NA	NA	0	47	343
Kuwait	0	128	222	300	294
Qatar	2	28	17	55	136
Neutral Zone	4	53	121	141	115
Algeria	21	10	3	3	103
UAE	21	267	252	339	42
Libya	30	17	34	72	40
Iran	15	143	316	148	40
Nigeria	231	653	720	665	207
Iraq	0	0	0	0	0
Venezuela	0	0	0	0	0
Indonesia	0	0	0	0	0
Total OPEC	1,797	3,332	4,358	5,226	4,539

Source: Petroleum Intelligence Weekly, September 3, 2007. a. Values for 2008 and 2010 are estimates.

Oil importing countries can create excess capacity in the world market by reducing their own demand, or augmenting their own oil production, if they hold oil reserves. While it is unlikely that any importing countries hold undeveloped reserves comparable to those in the larger OPEC nations, even more modest increases in production may reduce prices due to the price insensitive nature of oil demand. In this type of market, with a low degree of price sensitivity, relatively small changes in production might affect both current and future prices.

POLICY ISSUES

The increases in gasoline and oil prices have led to a number of congressional initiatives. Based on the number of bills introduced, a primary focus of the 110th Congress has been on curbing oil speculation activities. Proposed legislation has focused on: closing a variety of regulatory loopholes that have prevented regulatory authorities from exercising oversight, increasing regulatory resources, assigning emergency regulatory powers, and attempting to reduce the desirability of energy futures trading by increasing margin requirements, as well as data reporting.[15] Over three dozen pieces of proposed legislation were considered in the 110th Congress. It is possible, that because activities on the oil futures markets may have escaped regulatory scrutiny, strategies were undertaken that resulted in price manipulation.

Other policy approaches suggested to reduce gasoline prices included the suspension of the federal excise tax on gasoline. This tax is $0.1 84 cents per gallon, and the revenue generated is applied to the Highway Trust Fund. Concern was expressed as to whether the tax suspension would be passed on to consumers by refiners who collected the tax, and whether highway construction projects and the local employment they support would be adversely affected.[16]

The United States maintains a Strategic Petroleum Reserve of almost 750 million barrels of oil. Some have suggested that releasing oil from the reserve might reduce prices at the pump. Others felt that the amounts of oil that could be feasibly released were insufficient to affect gasoline prices, or that the OPEC nations might cut their output in response, cancelling any price effect.

Debate also has taken place concerning expansion of domestic supply, by drilling the restricted areas of the outer continental shelf, and other areas. This centered on areas currently excluded from oil development on the Atlantic and Pacific coasts, as well as some excluded areas in the Gulf of Mexico.[17] Also discussed was opening restricted areas in the Alaska National Wildlife Reserve to exploration and development. It is believed that producible oil deposits exist in these areas, but environmental concerns, the time lag required to complete development, as well as whether the deposits would be significant enough to affect prices in the current and future market, were critical factors in the debate.

CONCLUSION

Since the 1970s, the oil market has been both cyclic and volatile. Periods of high prices have been followed by price collapses. Up and down turns in price have been abrupt as well as drawn out. Policy proposals that assume that market behavior is predictable and can be projected into the future, are likely to be of limited effectiveness.

Additionally, there have been many goals set out for oil policy, some of them contradictory. For example, over the first half of 2008 there has been interest in lowering gasoline prices, reducing oil dependence, and reducing carbon emissions associated with global warming. Success in lowering gasoline prices would likely increase consumption, which likely would lead to increased oil imports, given the inability of the United States to increase production in the very short term. Increased consumption of petroleum products increases carbon emissions which could make the attainment of any desired emission target more difficult to obtain.

Market forces, both fundamental and financial, changed the direction of both the prices of oil and gasoline. In the five week period July 7, 2008 to August 11, 2008 national average gasoline prices fell from $4.165 to $3.864 per gallon, a decline of over 7%. NYMEX oil futures prices were $145.29 per barrel for WTI on July 3, 2008. On August 12, 2008, WTI was trading at about $113 per barrel, a decline of about 22%. These declines in prices were as unanticipated as the price increases from January through July. The oil market can be volatile and change direction quickly. That these observed price decreases come as Russia invaded Georgia, threatening oil pipeline shipments from Azerbaijan, unrest and violence continue to occur in Nigeria, and other de-stabilizing factors affect the market, makes their timing unanticipated.

Should these declines in price persist, and reverse most, or all, of the price increases that took place in the first half of 2008, the immediate pressure to control oil and gasoline prices through government policy might be reduced. Nevertheless, the market conditions that drove oil to record levels could quickly re-appear, if for example, U.S. economic growth picked up.

Policy making in a volatile market is difficult. It may also be hard for the industry to carry out a satisfactory investment plan that is consistent with policy objectives. The capital intensity and high costs of major oil projects require target prices to be attained to provide economic justification. In a sharply volatile market, the industry might not respond in the desired way to a policy measure. For example, policy measures that have the goals of increasing exploration and development of new oil fields, or expanding the capacity the oil refining sector as a way to mitigate high consumer prices, might not be undertaken by the industry if corporate planners view high prices as transitory, and lower prices are incorporated in the company's investment planning.

ENDNOTES

[1] Diesel and aviation fuel prices increased at least as much as gasoline prices.
[2] Table 1 represents U.S. regular gasoline, all formulations.
[3] Energy Information Administration, *This Week in Petroleum, Gasoline,* July 23, 2008.
[4] Federal Highway Administration, *Traffic Volume Trends,* May 2008.
[5] For example, total consumption demand in Japan and Europe declined between 2006 and 2007. This reduced demand translated into more available supply to satisfy Chinese demand.
[6] *BP Statistical Review of World Energy 2008*, June 2008, p. 6.

[7] See CRS Report RL3 1923, *Derivatives, Risk Management, and Policy in the Energy Markets*, by Robert Pirog, for more on the mechanics of the hedging process.
[8] See CRS Report RL34555, *Speculation and Energy Prices: Legislative Responses*, by Mark Jickling and Lynn J. Cunningham, for more on the focus and degree of oversight and regulation.
[9] David L. Crawford, "Oil Futures" Are a Phony Target, Philadelphia Daily News, August 4, 2008.
[10] Michael W. Masters, *Testimony before the Committee on Homeland Security and Governmental Affairs,* United States Senate, May 20, 2008.
[11] Ibid., p. 3.
[12] An increase in the value of the euro is the same as a decrease in the value of the dollar.
[13] CNN Special Report, *U.S. Gas: So Cheap it Hurts,* July 15, 2008.
[14] Excess capacity is sometimes referred to as "spare capacity."
[15] For more on these topics, see CRS Report RL34555, *Speculation and Energy Prices: Legislative Responses*, by Mark Jickling and Lynn J. Cunningham.
[16] For more on this topic, see CRS Report RL34475, *Transportation Fuel Taxes: Impacts of a Repeal or Moratorium*, by Robert Pirog and John W. Fischer.
[17] For more on this topic, see CRS Report RL33493, *Outer Continental Shelf: Debate Over Oil and Gas Leasing and Revenue Sharing*, by Marc Humphries.

In: Fuel Prices: Rhyme or Reason?
Editor: William P. Vestus

ISBN: 978-1-60692-842-4
© 2009 Nova Science Publishers, Inc.

Chapter 2

THE DISPARITY BETWEEN RETAIL GASOLINE AND DIESEL FUEL PRICES

Robert L. Bamberger and Robert Pirog

SUMMARY

Over time, gasoline has typically been more expensive than diesel fuel. However, their relative prices have now reversed. In mid-March of 2008, gasoline prices exceeded $3.39/gallon (gal) while diesel fuel prices were above $3.97/gal, a differential of almost $0.60/gal. This has prompted questions of why the historic gap between gasoline and on-highway diesel prices has widened so greatly and over such a relatively brief period of time.

Crude oil, when refined, produces a mix of products. Diesel fuel and home heating oil are derived from the portion of the barrel that produces what are termed "middle distillates." Another part of the barrel furnishes the feedstock for gasoline. Refiners process barrels of crude oil of differing quality, depending on the relative prices for oil of different qualities, and their available technology. Within technology-defined limits refiners can vary the proportions of middle distillate and gasoline production. Because the entire range of petroleum products derive from the same barrel, it is difficult to attribute general refining costs to any single product, making it also difficult to ascertain the relative cost proportions. The exception to this would be when the investment costs of changing product specifications to meet seasonal or environmental requirements can be measured.

A number of specific factors may be identified that have contributed to the shifting relative prices of gasoline and diesel fuel. It is important to recognize that the U.S. market for these fuels is part of a broader world market. World demand patterns are shifting as diesel fuel becomes a primary consumer transportation fuel in Europe and other parts of the world. World price differentials are transmitted to the U.S. market.

Other factors affecting diesel prices include refinery investment costs, as well as investment costs in the product distribution system to accommodate new specifications for diesel fuel that require lower allowable sulfur content; the seasonality of home heating oil demand, a similar product, which transmits the price effects of cold weather from the heating market to the on-highway diesel fuel market; world market effects that might affect the pricing and output mix decisions of refiners; and circumstances affecting the local market at point of purchase.

One other factor should be noted. The primary demand sectors for gasoline and diesel fuel are different in the United States. Gasoline is a mass consumer good and home heating oil an important regional and seasonal residential product, while diesel fuel is used in a wide variety of commercial and industrial applications. Diesel fuel is often part of the cost of delivering goods and providing services. As a consequence, demand for diesel fuel may be less elastic, and therefore, likelier to be passed on to consumers.

By mid-March of 2008, gasoline prices exceeded $3.39/gallon (gal) while diesel fuel prices were $3.97/gal, a differential of almost $0.60/gal. In mid-March of 2007, the relationship between the two fuels was the reverse: gasoline prices were higher than diesel prices.1 At that time, diesel prices were roughly $2.68/gal, while the average price of gasoline for all grades was $2.76 — more than $0.08 higher than the average price of on-highway diesel. Additionally, where gasoline prices in mid- March 2008 are roughly $0.63/gal higher than year-ago averages, diesel fuel prices have risen over $1 .29/gal over the same period.2 Over $0.60/gal of this increase has occurred since the beginning of 2008.3 This has prompted questions of why the historic gap between gasoline and on-highway diesel prices has widened so greatly and over such a relatively brief period of time. Because diesel fuel costs affect the cost of shipping by truck, price increases affect the delivered cost of most consumer goods purchased in the United States, contributing to the over-all level of price inflation.

This chapter provides background and identifies some of the likely factors and forces in world markets that may have contributed to the evolution of the relative prices of gasoline and diesel fuel over the past several years. Among these are strong international demand for diesel fuel; product mix decisions by refiners, and refinery investment to meet more stringent limits on the sulfur content of diesel fuel; the similarities between diesel fuel and home heating oil; and the effect on retail prices from local market conditions.

REFINING AND SUPPLY OF GASOLINE AND MIDDLE DISTILLATES

A barrel of crude oil is a composite of hydrocarbons of varying densities. The initial step in refining crude oil is to separate its heavier and lighter "fractions" by heating it. The lighter products are recovered at, or near the top of a distillation column where the temperature is lowest. The heavier fractions are recovered from the bottom where the heat is greatest. Gasoline is among the lighter components. Diesel fuel and home heating oil come from the portion of the barrel that is termed "middle distillates" because the feedstock for these fuels settle out roughly in the middle of the distillation tower.

Crude oil itself is of varying densities, as well as sulfur content, generally distinguished as "light" or "heavy," or high and low quality oil. Light crude will furnish a higher percentage of lighter products than heavy crude; additional processing can increase the yield of lighter products from the heavier end of the barrel, but will add to product costs.

Once distilled, gasoline and middle distillates are further processed "downstream," where the addition of blending components and other steps create the finished petroleum products that are released to markets. The typical yields from a barrel of crude oil of gasoline and middle distillates range, respectively, around 45- 47% and 25-27% depending on the time of year (see **Table 1**). Typically, refiners take some of their facilities offline for brief periods to perform maintenance and make seasonal adjustments to slightly favor the yield of gasoline or middle distillates. During the spring, refiners seek to build inventories of gasoline for the summer driving season. Conversely, production of home heating oil for the heating season is maximized beginning during the summer.

Gasoline consumption has been averaging 9.0 million barrels daily (mbd), while all distillate consumption is roughly 4.5 mbd. In 2007, U.S. imports of middle distillates averaged 348,000 b/d in 2007.[4] As is discussed later in this report, world demand for middle distillates has grown and added to the pressure on prices for middle distillate imports.

Table 1. Petroleum Products Produced from One Barrel of Oil Input to U.S. Refineries, 2006 (Product Gallons).

Product	Gallons
Finished Motor Gasoline	19.30
Distillate Fuel Oil	10.70
Kero-Type Jet Fuel	3.92
Petroleum Coke	2.24
Still Gas	1.87
Residual Fuel Oil	1.68
Liquefied Refinery/Petroleum Gas	1.58
Asphalt and Road Oil	1.33
Other Oils for Feedstocks	0.53
Naptha for Feedstocks	0.51
Lubricants	0.48
Other Products and Processing Gain	3.31

Source: Energy Information Administration, Department of Energy.

The typical product yield from a barrel of crude oil is shown in **Table 1**.

As is also noted later in this report, some of the sharp runup in on-highway diesel fuel prices in recent months likely stems from the close similarity between diesel fuel and residential home heating oil. Both, as has been noted, are middle distillates and, to some extent, in competition with one another. Home heating oil and transportation diesel are chemically identical, but in the refinery they are processed in slightly different ways for their respective purposes. In addition to having specified regulations and taxes, transportation diesel has a low sulfur standard, meaning that it must contain 0.05% sulfur or less. Home heating oil is required by law to contain not more than 0.5% sulfur content, but due to unintentional mixing of transportation diesel and home heating oil at the refinery, the sulfur content of home heating oil usually hovers around 0.2%.

Table 2 shows recent demand for products that fall within the parameters of distillate fuels. In 2006, diesel fuel represented nearly 63% of distillate sales while residential home heating oil was 8% of sales. This is compared with 58% and 10.8%, respectively, in the year 2002.

Table 2 also suggests that the distillate fuel market in the United States is not a growth market. The total demand for distillates was no higher in 2006 than in 2003, and less than 2% higher than the weak demand year of 2002. On-highway diesel was the only sector that showed continuous growth over the period. Other sectors, like residential and commercial, suggest seasonality related to the weather. Some sectors, like vessel bunkering, electric power, and military showed declining demand.

Table 2. Sales of Distillate Fuel Oil by End Use in the United States: 2002-2006 (thousand gallons).

Energy Use	Distillate Fuel Oil				
	2002	2003	2004	2005	2006
U.S. Total	59,342,633	63,854,776	62,257,934	63,164,569	62,192,027
Residential	6,376,653	6,927,070	6,644,939	6,154,461	4,984,826
Commercial	3,293,387	3,686,537	3,383,061	3,224,216	2,808,786
Industrial	2,384,383	2,394,445	2,326,604	2,459,711	2,463,676
Oil Company	770,682	513,511	472,920	472,922	636,788
Farm	3,418,452	3,200,809	3,189,014	3,215,819	3,261,345
Electric Power	750,557	1,147,727	823,380	906,976	656,355
Railroad	3,245,482	3,656,657	3,047,491	3,447,630	3,552,430
Vessel Bunkering	2,078,921	2,216,921	2,139,643	2,005,564	1,903,138
On-Highway Diesel	34,308,885	37,103,563	37,125,239	38,053,129	39,118,301
Military	357,359	415,702	358,682	268,553	327,827
Off-Highway	2,357,872	2,591,833	2,746,960	2,955,589	2,478,554

Source: Department of Energy, Energy Information Administration.

Differing sectoral demand patterns within the same product group makes it likely that, in pricing terms, those sectors with the relatively strongest demand patterns might be charged prices which help to offset the lower returns that might be earned in sectors with weaker demand. For example, since all of the distillates are joint products of the refining process, all must find a market. However, if one segment of the market, say use in electric power generation, is relatively weak and declining, and another, on-highway use is increasing, it is likely that electric power distillates may be sold at a discount, while on-highway distillates may be sold at a premium.[5]

DIESEL AND GASOLINE PRICES

The retail prices of gasoline and diesel fuel have four major components: the price of the crude feedstock; federal and state taxes; the cost of refining, reflected in what is referenced as the "refiner margin"; and the costs of distribution (transportation) and marketing. As the price of crude rises or fluctuates, along with any demand pressures, the relative percentage share of these components of retail price will shift. The observed drop in the share represented by state and federal taxes — values that are constants over the period shown below — is a reflection of the significant change in the retail sales price for gasoline and diesel fuel. These percentages for the last year for both gasoline and diesel fuel are set out in **Tables 3** and **4**, and depicted as graphs in **Figures 1** and **2**.

Table 3. Components of Retail Gasoline Price: January 2007-January 2008.

Mo/Year	Retail Price (Cents per gallon)	Refining (percentage)	Distribution and Marketing (percentage)	Taxes (percentage)	Crude Oil (percentage)
Jan-07	224.0	10.6	15.2	20.3	53.9
Feb-07	227.8	18.0	5.8	20.0	56.3
Mar-07	256.3	23.6	8.5	15.5	52.3
Apr-07	284.5	28.1	7.6	14.0	50.3
May-07	314.6	27.9	13.3	12.7	46.1
June-07	305.6	22.7	13.7	13.0	50.5
Jul-07	296.5	18.4	11.4	13.4	56.8
Aug-07	278.6	13.5	11.8	14.3	60.4
Sep-07	280.3	12.8	8.6	14.2	64.3
Oct-07	280.3	10.1	8.1	14.2	67.6
Nov-07	308.0	10.0	8.7	13.0	68.3
Dec-07	301.8	8.1	10.5	13.2	68.1
Jan-08	304.3	7.8	11.1	13.1	67.9

Source: Energy Information Administration, Department of Energy. [http://tonto.eia.doe.gov/oog/info/gdu/gaspump.html]

The Disparity Between Retail Gasoline and Diesel Fuel Prices 23

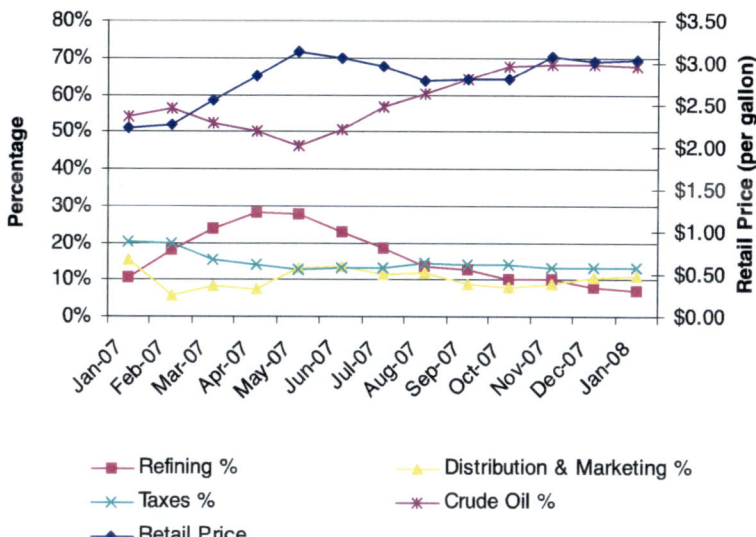

Source: Energy Information Administration, Department of Energy. [http://tonto.eia. doe. gov/oog/ info/gdu/gaspump.html] Adapted by CRS.

Figure 1. Components of Retail Gasoline Price: January 2007-January 2008.

Table 4. Components of Retail Diesel Price: January 2007-January 2008.

Mo/Year	Retail Price (Cents per gallon)	Refining (percentage)	Distribution and Marketing (percentage)	Taxes (percentage)	Crude Oil (percentage)
Feb-07	248.8	21.5	5.8	21.1	51.5
Mar-07	266.7	23.6	8.7	17.4	50.3
Apr-07	283.4	23.4	9.7	16.4	50.5
May-07	279.6	22.0	9.6	16.6	51.9
June-07	280.8	21.0	7.5	16.5	54.9
Jul-07	286.8	18.3	6.8	16.2	58.7
Aug-07	286.9	16.0	9.2	16.1	58.7
Sep-07	295.3	17.4	5.9	15.7	61.0
Oct-07	307.5	16.7	6.6	15.1	61.6
Nov-07	339.6	17.1	7.3	13.7	61.9
Dec-07	334.1	15.4	9.1	13.9	61.6
Jan-08	330.8	14.6	8.9	14.0	62.5

Source: Energy Information Administration, Department of Energy.

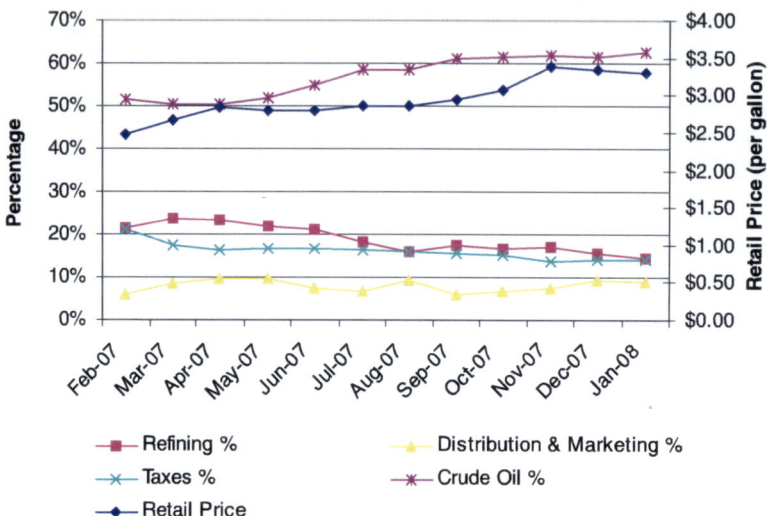

[http://tonto.eia.doe.gov/oog/info/gdu/dieselpump.html]
Source: Energy Information Administration, Department of Energy. [http://tonto.eia. doe. gov/oog/ info/gdu/dieselpump.html] Adapted by CRS.

Figure 2. Components of Retail Diesel Price: January 2007-January 2008.

Tables 3 and **4** suggest that the reason for the shift in the relative prices of gasoline and diesel fuel cannot be easily be identified through cost growth at any particular stage of the production process. However, part of the explanation may be in the behavior of refining as a percentage of price. The decline in refining cost in gasoline has been greater than the decline in refining cost in diesel. This pattern suggests that the ability of refiners to pass through cost increases to the consumer is stronger in diesel than in gasoline, and that there are significant recoverable costs that have been added in diesel refining. Both may play a role. As is discussed later in this report, mandated refinery investments have been required in diesel fuel refining to meet new product specifications.

On the demand side, the second half of 2007 and early 2008 have been characterized by record-setting crude oil prices. Gasoline prices lagged the increase in crude oil prices, leading to shrinking refiner margins and profitability. Possibly, because diesel fuel is an intermediary product in commercial use, and as such can be expected to be passed through to final consumers, refining costs as a percentage of cost remained stronger.

It is expressly because refiners often absorb the initial increases in crude prices that some are predicting that, if crude prices remain roughly in their current range or go higher, further price increases in all highway fuels are likely. How steep these increases may prove to be will depend very critically on the demand response to the price of motor fuels. Gasoline demand is recently observed to be relatively flat, and stocks of gasoline are unseasonably high. However, with the start of the summer driving season still some weeks off, any prediction about the price, supply, and demand for gasoline (and diesel fuel) during the summer of 2008 would be conjecture at best.

Factors Influencing Gasoline-Diesel Fuel Price Differential

A variety of factors, some cyclical, and some structural, have likely contributed to the break-down of the traditional pattern of relative prices between gasoline and diesel fuel. These are identified and described in turn.

World Market Balance

Growing petroleum product demand, including demand for diesel fuel, in China, Europe, and the United States has put pressure on the ability of refineries to meet production requirements. Demand growth in China is primarily tied to the level of economic growth, expanding both industrial and consumer demand. While the overall growth in petroleum demand in Europe has not been high, demand for diesel fuel over gasoline has increased. The European automobile and light truck fleet has moved in the direction of diesel fuel. In the United States, the demand for gasoline has continued to increase. Even though crude oil prices have risen since 2004, demand for gasoline in the United States over the same period continued to increase.

In a world market where the major producers sell their products in virtually every geographic and product segment, price effects will have a tendency to move from one part of the market to another. If strong demand for diesel fuel exists in Europe and places upward pressure on prices, the effect is also likely to be felt in the U.S. market. Even if it were possible to wall off the U.S. market from higher prices, it is unlikely that it would be helpful. If a price spread between gasoline and diesel, greater than the cost of shipping, develops between Europe and the United States, a major oil company might be inclined to draw diesel fuel from the U.S. market and sell it in Europe to earn a greater profit. The potential for transactions of this type transmit price increases from one geographic market to others, even if the trade flow does not occur.[6]

U.S. imports of diesel fuel have been in the 200 to 400 thousand barrels per day range since 2004. If the import price of diesel fuel exceeds the domestic price of the same fuel, and given that in the market all product is sold at the same price, all prices will rise to the level of the higher cost imported fuel. In terms of the example cited above, tight demand and supply conditions in the European diesel fuel market are transmitted to the U.S. markets as prices tend to equalize.

Refinery Output

U.S. refinery utilization rates in recent years have been high, generally at or near 90%, reflecting strong domestic demand for most petroleum products. The product mix has generally been optimized to produce a maximum amount of gasoline. This could be true even in times when the price of diesel fuel is above the price of gasoline. Record profit levels in the oil industry have increased public scrutiny of oil company operations. Although prices of all petroleum products have been high, and market conditions tight, physical shortages of transportation fuels have not been generally observed. If the general motoring public had to confront high gasoline prices at the same time that physical shortages were developing, the pressure to tax or regulate oil company profits and product prices might grow. As a result, it may be that a major priority of the oil companies supplying the U.S. market is to avoid shortages. To avoid shortages, the U.S. imports gasoline and gasoline blending components. These imports now generally exceed 1 million barrels per day, augmenting domestic gasoline production, and avoiding the likelihood of physical shortages of gasoline.

A possible result of maximizing gasoline output at the refinery may be to make the supply of diesel fuel relatively less available when compared to any particular level of demand, resulting in stronger upward pressures on diesel fuel prices compared to gasoline prices. In this way, even though all petroleum product prices are rising due to the increasing price of crude oil, the relative prices of diesel fuel and gasoline could shift because of an emphasis on gasoline production.

Sulfur Content

The Environmental Protection Agency (EPA) in 2001 promulgated new rules concerning the sulfur content of diesel fuel that began to go into effect in 2006. Ultra low sulfur diesel (ULSD) contains 15 parts per million of sulfur, compared to 500 parts per million or more in uncontrolled diesel fuel. Refineries were to begin producing 80% of their output of diesel fuel as ULSD in June 2006, with availability at fuel outlets for on-highway use by October 2006. Because the sulfur content is measured at the pump according to EPA regulation, special transportation and distribution systems were also needed to avoid fuel contamination. Use of reduced sulfur diesel for off-highway purposes began in 2007, with full implementation of ULSD by 2010.

The American Petroleum Institute estimated that over $8 billion have been spent by refiners to acquire and implement refinery processes for sulfur removal. In addition, hundreds of millions of dollars have been spent to upgrade transportation and distribution systems. These investment costs to meet federal regulation are likely to be passed on to consumers in the form of higher diesel fuel prices.[7] These investment costs increase the refinery cost component of diesel fuel, and if the refiners allocate costs specifically to the cost-generating product, diesel prices should rise relative to gasoline prices.

Heating Oil/Seasonality

Home heating oil and diesel fuel are essentially the same product from the refining point of view, and as such, their prices are related in the market. As a result, peaking demand for home heating oil in cold months can have an effect on the price of diesel fuel.

For parts of the United States, the winter of 2007-2008 was colder than usual.[8] Heating oil prices reached a record price of $3.55 per gallon for the week ending March 3, 2008. This record price represented an increase of almost 9 cents from the previous week. These prices represented higher than a year-ago prices for the 22nd consecutive week this heating season. Heating oil demand and high prices have likely contributed to the increases observed in diesel fuel prices.

In addition, the linkages between the domestic diesel fuel market and international markets suggest that cold weather which increases heating oil demand anywhere in the world is likely to contribute to higher heating oil and diesel fuel prices in the United States.

Pricing Practices

In a market economy, sellers of a commodity may set prices at whatever level they think the market will bear. Consumers respond by adjusting their level of purchases. If the consumer's demand is inelastic, or insensitive to price, then sellers have an incentive to charge higher prices. Transportation demand, and hence the demand for fuels including gasoline and diesel fuel, is thought to be relatively price insensitive in the short term. In addition, since diesel fuel is used for mostly business purposes in the United States, it may be treated as an intermediate good; one that is a cost component of a production process leading to some final consumer good or service. As such, any increases in diesel fuel costs are likely to be passed on to the ultimate consumers. If costs can be passed on through a pricing process, there is little need for those who use the product to make adjustments as a result of higher costs.

Although gasoline and diesel fuels are joint products of the refining process, refining companies have the right to apportion the costs of production to segments of the product mix in whatever blend they choose. Refiners may choose to change relative prices within the product mix to take advantage of demand conditions, to alter the composition of demand to match available supply, or simply as a strategy to increase shareholder value.

CONCLUSION

On the basis of the market dynamics described in this report, the future price path of highway fuels and the relative disparity between the price of gasoline and diesel fuel cannot be predicted with any confidence. At this time, the price support for diesel fuel is primarily demand-driven, with the United States competing for world supply to supplement domestic production of middle distillates with product imports.

It could be anticipated that, at some point, the price of a fuel could reach a level where there is some demand response. It is unclear what these price points may be. However, owing to the primary use of diesel fuel in the commercial sector for the delivery of goods and some services, demand for diesel is likely to be less elastic because, as has been noted, those costs will be passed on to consumers. Demand outside the United States may also prove to be less elastic. A supply response could ameliorate prices somewhat, but any supply response is bounded by the nature of crude oil and refinery investment.

ENDNOTES

[1] Gasoline prices exceeded diesel fuel prices for every year between 1995 and 2004. For 2005 through 2007 diesel prices were, on average, above gasoline prices. However, the quarterly results were mixed; gasoline prices exceeded diesel prices in some quarters. See the Energy Information Administration, *Annual Energy Review, 2006*, Table 5.24, and comparative gasoline and diesel price data at [http://www.eia.doe.gov].

[2] The most recent weekly prices for gasoline and diesel fuel at the pump are reported by the Energy Information Administration, Department of Energy, at these sites: [http://www.eia.doe.gov/oil_gas/petroleum/ data_publications/wrgp/mogas_home_page.html, and] [http://tonto.eia. doe.gov/oog/info/wohdp/diesel.asp].

[3] See [http://tonto.eia.doe.gov/oog/info/wohdp/List_Serve_report_All.txt].

[4] See U.S. Department of Energy. *Petroleum Supply Monthly*, Table 36. Year-To-Date Imports of Crude Oil and Petroleum Products by PAD District, January-December 2007. [http://www.eia.doe.gov/pub/oil_gas/petroleum/data_publications/petroleum_supply_monthly/current/pdf/table36.pdf].

[5] Economists identify joint products in production as products which must be produced together. The nature of the technology, or the raw materials, are such that if one product is produced, each of the products in the joint product group must be produced, even though the demand conditions for each product may be very different.

[6] Economists relate this kind of analysis to opportunity costs. Opportunity costs are the value of a good or service in its next best alternative use. Prices are generally at, or above, their opportunity values, but it is unlikely they will trade below that value.

[7] American Petroleum Institute, *Diesel Fuels,* at [http://www.api.org].

[8] See map of accumulated heating degree days for the United States, November 2007 through March 11, 2008 at [http://www.cpc.ncep. noaa.gov/products/predictions/experimental/ddtest/sdhdd.glf].

Current market conditions, however, may limit, or even prevent, an observed reduction in prices to end-users. Rising crude oil prices have resulted from a continued growth in world demand that has put pressure on available world supplies, resulting in diminished world excess oil production capacity and generally tight inventories. These conditions have persisted in the oil market since 2003. In addition, some analysts have identified other factors, including speculation on oil futures markets, inflation hedging, political disruption, and mismanagement by national oil companies, as contributing to high oil prices. Rising crude oil prices may continue to push up the price of gasoline, offsetting any reduction in the fuel taxes.

With respect to refined petroleum products, especially gasoline and diesel fuel, demand in the United States has exceeded domestic refining capacity, even though the refining industry has run at high capacity utilization rates. As a result, about 1 million barrels per day of finished gasoline and gasoline blending stocks are imported to meet U.S. demand.[7] The combination of limited capacity and high capacity utilization rates greatly limits the ability of the refining industry to increase the availability of refined petroleum products. Under these conditions, most experts would likely conclude that it is probable that little, if any, of a tax cut due to a temporary suspension of the fuel excise taxes would be passed forward to final consumers.

Notwithstanding generally high profits in the oil industry, profit margins in the the refining sector have been much lower. Four (Shell, BP, Chevron, and Marathon) of the six (ExxonMobil, ConocoPhillips) largest integrated oil companies experienced declining income growth in downstream operations in 2007.[8] Reduced profits in the refining sector might provide a further incentive for the firms not to pass tax cuts on to consumers, especially in light of the temporary nature of the tax cut.

Moreover, because 18.4 cents is small in relation to current end user prices for transportation fuels, about 5%, most experts would likely conclude that even a full pass-through of a suspension or repeal would have little effect on end user prices. In an environment of rapidly increasing prices, even a full pass-through might only offset the price increases emanating from market pressures. In that case, consumers might continue to see increasing prices at the pump, even though the tax cut was made fully available to them. If lesser amounts than the full tax cut were passed on to consumers, market-generated price increases might make the effect of the tax cut inconsequential.

Proposals to suspend the full amount of only the gasoline, or diesel, fuel tax could be expected to have more complicated market effects. Although temporary, singling out one fuel for tax relief could change the relative price structure among refined products and introduce incentives to change the proportions of products derived from each barrel of crude oil, all of which would be affected by the seasonality of demand. Also, it is possible that refiners would apply some of a particular product tax cut to the prices of one or more other refined products. Thus, it is uncertain not only that measures that give excise tax relief will result in reductions in prices to end users, but that any reductions will apply to the product given tax relief.

If the tax cut is passed on to users of gasoline, and prices are reduced, it is likely that consumption will increase. An increased consumption of gasoline will lead to an increased demand for oil that will likely increase its price and lead to further increases in the price of gasoline. In addition, increased demand for gasoline in the United States is likely to be met by an increase in imported gasoline, increasing U.S. dependence on foreign supplies. Increased consumption of gasoline will also increase carbon emissions.

Also, although it may not be likely, some states experiencing budget pressure could substitute increased state fuel taxes for the reduced federal tax. In that case, if the refiners did not pass through the full tax cut to consumers, and if a state raised its tax by an amount equal to the proposed reduction in the federal tax, the net tax effect on consumers might increase. This response could result from a decrease in funding from the Highway Trust Fund. A few states have provisions that automatically increase their taxes to some extent when federal taxes fall below a certain level.

SURFACE TRANSPORTATION PROGRAM EFFECTS

Federal funding for surface transportation is closely linked to the revenue stream provided by the Highway Trust Fund. In actuality, the trust fund consists of two separate accounts — highways and mass transit. In common usage, the term Highway Trust Fund normally refers to the highway account. As mentioned earlier, the primary revenue sources for these accounts are the 18.4-cent-per-gallon tax on gasoline and a 24.4-cent-per-gallon tax on diesel fuel. Although there are other sources of revenue for the trust fund, these fuel taxes provide about 90% of the income to the funds. Of these amounts, the transit account receives 2.86 cents per gallon, and 0.1 cent per gallon is reserved for an unrelated leaking underground storage tank (LUST) fund. Over the almost 50-year life of the trust fund, there have been several increases in the level of taxation. The last increase in the fuel tax occurred in 1993.[9]

Growth in the trust fund revenue stream over the last five decades has remained reliable largely because of continued growth in auto and truck registrations, which, combined with increased auto and truck use, has resulted in a relatively constant annual increase in national fuel consumption. Growth in fuel consumption has also been enabled by changes in the makeup of the national vehicular fleet. This was especially the case in the 1 990s, when consumer demand for SUVs and light trucks led to an even higher level of growth in fuel usage.

Although driving has not yet decreased significantly as a result of $3.50-plus-per-gallon fuel costs, it is believed that these price levels are affecting vehicle purchase decisions and other use-related decisions.[10] For example, fuel-efficient vehicles such as hybrids are selling well, and less fuel-efficient vehicles are selling at a much reduced rate. This situation will, if it continues, potentially reduce longterm growth in fuel tax revenue. Other fuel cost-related trends may develop over time. For example, a significant shift to alcohol-based fuels would put pressure on the overall federal budget because of federal subsidies afforded these fuels.

Federal Surface Transportation Program Revenue Issues

Federal surface transportation programs are currently authorized by the Safe, Accountable, Flexible, Efficient Transportation Equity Act — A Legacy for Users (SAFETEA-LU or SAFETEA) (P.L. 109-59). This act reauthorized federal surface transportation programs through the end of FY2009. The act provided $286.4 billion for a six-year authorization period (in actuality, the act provided $244.1 billion for the five years remaining in the authorization at the time of passage).

Of immediate concern to the transportation community is the fact that the federal-aid highway program is currently spending more on highways on an annual basis than the Highway Trust Fund is receiving annually in new revenues. Although this trend began prior to passage of SAFETEA, the funding levels provided by the act can only be met by spending down the cash balances in the highway and transit accounts of the trust fund. The authors of SAFETEA believed that the existing cash balances in the trust fund, when combined with new revenues, were sufficient to carry the program through the FY2009 authorization period. There is now widespread agreement in the transportation community that this was an optimistic prediction. Both the Office of Management and Budget (OMB) and the Congressional Budget Office (CBO) believe that the cash balances in the highway account will be negative prior to the end of FY2009.[11] A fuel tax repeal or moratorium without a concomitant increase in revenues from some other source, such as the Treasury general fund, could hasten and otherwise exacerbate this situation.

At the time of this writing, no CBO, OMB, or Joint Tax Committee estimate of the amount of revenue that would be forgone by a gas tax holiday is available. A rough estimate of how much a tax holiday could cost the trust fund can be derived as follows. The Bush Administration budget proposal for FY2009 estimates that the highway account of the trust fund will collect revenues of $34.19 billion in FY2008.[12] The transit account will collect an estimated $5.01 billion for the same period. Total estimated revenue collection for the fund for FY2008 would, therefore, be $39.2 billion. As mentioned earlier, approximately 90% of trust fund revenues derive from fuel taxes. Therefore, approximately $35.28 billion of FY2008 revenues would be from fuel taxes. If annual fuel sales are evenly divided over 12 months, each month that the fuel tax might be suspended would result in $2.94 billion in lost revenue. A three-month suspension, similar to those proposed by S. 2890 and S. 2971, equates to $8.8 billion under these assumptions. Since the suspension period proposed in this legislation represents the peak summer driving period, it is possible that the amount forgone could be slightly higher.

Because of the manner in which federal transportation programs operate, the effects of these reductions would not become apparent in highway project spending immediately. Federal transportation programs are reimbursable programs, meaning that actual outlays only occur after work has been completed at the state or local level. The FY2009 budget predicts that the Highway Trust Fund would have an endof-year FY2008 balance of $3 billion. If alternative funding were not provided to make up for the fuel tax holiday lost revenues, it is likely, therefore, that the Highway Trust Fund will have a negative balance at some point prior to the end of FY2008. This would not be the case for the transit account which is operating with a larger unexpended balance. At some point, the Federal Highway Administration (FHWA) would have to defer outlays for highway project reimbursement, pending the accrual of revenues to the highway account, which would be minimal until the reinstatement of the fuel taxes. It is also likely that FHWA would deem it prudent to reduce or otherwise delay the award of new obligation authority to the states (this is the authority that allows states to enter into contracts for new projects).

The programs authorized by SAFETEA are due for reauthorization at the end of FY2009. Any disruption in surface transportation funding could hasten the reauthorization debate, which was expected to take place in the first session of the

111th Congress. Reauthorization is expected to include a significant discussion about the size of the overall federal commitment to surface transportation funding and the adequacy of existing revenues. A suspension of the fuel taxes, assuming that they would be reinstated at the end of the holiday period, could raise problematic questions about the supposed sanctity of the five-decade-old link between fuel taxes and surface transportation funding. This would especially be the case if surface transportation programs were to become ever more dependent on Treasury general funds.

ENDNOTES

[1] Cave, Damien. "States Get In on Calls for a Gas Tax Holiday." *The New York Times*, May 6, 2008.

[2] The summer driving season, from Memorial Day until Labor Day, is believed to result in an increase in the demand for gasoline due to holiday and vacation travel as well as increased driving for personal errands and commercial activity due to longer benign weather daylight hours.

[3] CRS has estimated fuel costs as at least 15% of total operating costs, based upon data from trucking company annual reports, the American Trucking Associations, Global Insight, and the Energy Information Administration (U.S. Department of Energy).

[4] Economists measure the responsiveness of the supply by estimating the value of the "supply elasticity."

[5] Economists identify this situation as an infinite elasticity of supply.

[6] Depending on the particular circumstances, federal transportation fuel excise taxes are levied on and remitted by the refiner, terminal operator, or importer.

[7] Energy Information Administration, *Weekly Imports and Exports*, available at [http://www.eia.doe.gov].

[8] Downstream operations include refining, distribution, and marketing.

[9] A significant portion of the 1993 increase was initially deposited in the Treasury general funds for deficit reduction purposes. These funds were redirected to the Highway Trust Fund beginning in FY1998.

[10] Vlasic, Bill. "As Gas Costs Soar, Buyers Are Flocking to Small Cars." *The New York Times*, May 2, 2008.

[11] There is widespread concurrence within the transportation finance community that the trust funds are in financial difficulty and that the highway account balances will be negative before the end of FY2009. The primary item of continuing debate is the point during the fiscal year at which this might occur. For a discussion of the overall financing issue, see National Surface Transportation Infrastructure Financing Commission, Interim Report, *The Path Forward: Funding and Financing Our Surface Transportation System*. Washington. February 2008. p. 11.

[12] Office of Management and Budget. Budget of the U.S. Government, Fiscal Year 2009, Appendix. Washington, 2008, p. 888 and p. 917.

In: Fuel Prices: Rhyme or Reason? ISBN: 978-1-60692-842-4
Editor: William P. Vestus © 2009 Nova Science Publishers, Inc.

Chapter 4

MOTOR FUELS: STAKEHOLDER VIEWS ON COMPENSATING FOR THE EFFECTS OF GASOLINE TEMPERATURE ON VOLUME AT THE PUMP

United States Government Accountability Office

WHY GAO DID THIS STUDY

The volume, but not the energy content, of hydrocarbon fuels, such as gasoline and diesel, varies in response to changes in temperature. Thus, because of expansion, the energy content per gallon of 90 degree fuel is less than that of 60 degree fuel. States and localities adopt and enforce weights and measures regulations, often using the model regulatory standards published by the National Institute of Standards and Technology (NIST). Although technology now exists to compensate for the effects of temperature on gas volume, the costs of doing so at the retail level have become the subject of much debate among weights and measures officials, consumer groups, and representatives of the petroleum and fuel marketing industries.

GAO was asked to provide information on (1) the views of U.S. stakeholders on the costs to implement automatic temperature compensation, (2) the views of U.S. stakeholders on who would bear these costs, and (3) the reasons some state and national governments have adopted or rejected automatic temperature compensation. To do this work, GAO reviewed NIST and other documents and congressional testimony; interviewed stakeholders from 3 federal agencies, 17 states, and 15 groups representing a variety of interests, including consumers, truck drivers, and the oil and gas industry; and interviewed officials in 5 other nations.

Various stakeholders and officials provided technical and other comments, which were incorporated in the chapter as appropriate.

WHAT GAO FOUND

The costs to implement automatic temperature compensation are unclear. Most stakeholders said that implementing automatic temperature compensation for retail sales would involve the cost to purchase, install, and inspect new equipment on pumps, as well as costs to educate consumers about the change. Some stakeholders said the costs to adopt automatic temperature compensation ranged from $1,300 to $3,000 per pump, but none had estimated the total costs nationwide, in part because complete data are not available. Estimates of the cost to inspect the new equipment varied. Officials in a small number of states said inspection times would increase by 20 to 50 percent, while officials in three other states said the costs would not be significant. No stakeholders had developed estimates of the costs to educate consumers.

Stakeholders differ on whether retailers, consumers, or both would ultimately bear the costs of implementing automatic temperature compensation at the retail level. Some stakeholders, including state officials and industry representatives, said that the costs would be passed on to consumers through higher prices for fuel or other goods sold at retail stations. Others, such as consumer groups, said that retailers and consumers would share the costs and benefits. That is, some retailers could use funds they receive from major oil companies for remodeling to pay for the equipment. These stakeholders also said the benefits include consistent energy content for consumers and improved inventory management for retailers. Stakeholder views were largely based on professional judgment and general economic theory rather than on studies or other data, and most stakeholders said that a comprehensive cost- benefit analysis would provide policymakers with important information.

Governments that have adopted or permitted automatic temperature compensation for retail fuel sales cited improved measurement accuracy and greater equity between retailers and consumers as reasons for making the change; those that have prohibited it largely cited concerns that the costs would outweigh the benefits. Hawaii adopted temperature compensation more than 26 years ago because it provided purchasing equity for the industry and consumers. In 2008, Belgium mandated temperature compensation to help ensure more consistent energy content for consumers. Canadian officials cited improved measurement equity and accuracy as reasons for allowing retailers to sell temperature-compensated fuel in the early 1990s. In the United States, officials from eight states that have laws or regulations that prohibit automatic temperature compensation said the decision should be based on an analysis of the costs and benefits, with some expressing concern that the costs would outweigh the benefits. None of the governments that have adopted automatic temperature compensation have studied its impact.

LETTER

September 25, 2008

The Honorable Bart Gordon
Chairman
Committee on Science and Technology House of Representatives

Dear Mr. Chairman:

Consumers and businesses alike are concerned about the steep rise in fuel prices in recent years. Because the volume of hydrocarbon fuels, such as gasoline and diesel,[1] varies in response to changes in temperature, some are concerned about the potential impact of temperature-related changes in volume on the amount they pay. More specifically, the volume of gasoline expands or contracts by 1 percent for each 15 degree increase or decrease in temperature, while the energy content of gasoline remains the same. For example, 10 gallons of gasoline at 60 degrees Fahrenheit (F) expands to 10.2 gallons of gasoline at 90 degrees F but maintains the same total energy content.[2] As a result, the average energy content per gallon of the 90 degree fuel will be less than that of the 60 degree fuel. In the United States, wholesale fuel transactions are routinely adjusted for temperature- related changes in volume. However, at the retail level, gasoline and diesel are sold by volume—specifically, 231 cubic inches per gallon—without regard to temperature, leading some to believe that the retail price of a gallon of fuel may not reflect its true value. Advances in measurement technology have allowed the development of devices that can automatically compensate for the effects of temperature on volume when dispensing fuel at retail gas pumps.[3]

While some argue that extending temperature compensation to the retail level could provide greater transparency in fuel prices, others contend that the cost to upgrade existing equipment could be substantial and impose economic hardship on retailers.

The National Conference on Weights and Measures (NCWM), a consensus-building organization composed of state and local regulatory officials and other interested parties, has discussed whether to adopt standards for temperature compensation of gasoline and diesel for over 30 years, most recently at its meeting in July 2008. NCWM plays a key role in the debate because states adopt and enforce weights and measures regulations.

NCWM receives technical guidance on this and other matters from the Office of Weights and Measures in the Department of Commerce's National Institute of Standards and Technology (NIST). In partnership with NIST, NCWM develops model regulatory standards that are available for adoption and enforcement by state or local weights and measures authorities. NIST publishes these standards in various handbooks, and any proposed changes to these handbooks are considered by NCWM.

Since 2000, NCWM has considered various proposals related to automatic temperature compensation, including proposals in 2007 and 2008 to adopt model regulatory standards that states could use to implement temperature compensation in retail sales of gasoline and diesel. Neither of the proposed model standards has been adopted. In addition to the deliberations of NCWM, the Congress has held hearings on the issue, and federal legislation has been proposed to require the use of temperature compensation in retail transactions. However, the economic implications of temperature-induced changes in the volume of motor fuels on the price of gasoline and diesel remains a topic of considerable debate, and the issue continues to elicit strong opinions, both for and against, from parties such as petroleum marketers, retailers, independent truckers, fleet owners, and consumer advocates.

In the context of this debate, you asked us to provide information on (1) the views of U.S. stakeholders[4] on the costs to implement automatic temperature compensation, (2) the views of U.S. stakeholders on who would bear these costs, and (3) the reasons some state and national governments have adopted or rejected automatic temperature compensation. For each of these issues, we agreed to report on the support, such as studies or data, that stakeholders use for their views.

To obtain information from U.S. stakeholders on the costs to implement automatic temperature compensation and who would bear those costs, we reviewed NCWM documents and congressional testimony and performed a literature search to identify relevant documents and stakeholders likely to have a view on the implementation of automatic temperature compensation in the United States. To identify additional stakeholders, we asked each stakeholder we interviewed for recommendations of knowledgeable other entities and selected for interviews those who would provide us with a broad and balanced range of perspectives on temperature compensation of gasoline and diesel. We used a standard set of questions to interview each of these individuals to ensure we consistently discussed each aspect of automatic temperature compensation. Specifically, we interviewed representatives of two consumer advocacy groups, five fleet owners and operators, a former NIST official, and officials at seven organizations that represent independent truck drivers, the oil and gas industry, independent petroleum marketers, convenience store and truck stop owners, and the trucking industry. To obtain views from governments that have adopted or rejected temperature compensation, we contacted officials in 16 states that have taken specific steps to adopt or prohibit automatic temperature compensation. We also contacted officials in California who are conducting a cost-benefit analysis of temperature compensation. In addition, we contacted officials from Australia, Belgium, Canada, the United Kingdom, and a European weights and measures organization because literature and interviews indicated these governments had adopted or had considered implementing automatic temperature compensation. We also interviewed officials from the Environmental Protection Agency (EPA), the Federal Trade Commission (FTC), and NIST because these agencies help oversee the marketplace generally or oversee aspects of the retail petroleum industry. See appendix I for a more detailed description of the methodology we employed.

We conducted our work from March 2008 to September 2008, in accordance with generally accepted government auditing standards. Those standards require that we plan and perform the audit to obtain sufficient, appropriate evidence to provide a reasonable basis for our findings and conclusions based on our audit objectives. We believe that the evidence obtained provides a reasonable basis for the information we present for each of our audit objectives.

RESULTS IN BRIEF

The costs to implement automatic temperature compensation are unclear. Stakeholders said that implementing automatic temperature compensation for retail fuel sales would involve costs to purchase, install, and inspect new equipment on fuel pumps, as well as costs to educate consumers about the change. Although some stakeholders had limited estimates for costs associated with the adoption of automatic temperature compensation, ranging from $1,300 to $3,000 per pump for the costs to purchase and install automatic temperature compensation equipment, none had estimated the total magnitude of these costs nationwide. These estimates from stakeholders were generally consistent with information we obtained from equipment manufacturers. Specifically, costs ranged from $900 to $1,800 to buy a kit to retrofit an existing pump and $200 to install the kit. Stakeholders said the costs to adopt temperature compensation could be affected by such factors as whether the investment to adopt the devices occurred immediately or more gradually to accommodate routine replacement decisions by retailers. A small number of stakeholders said estimates of the magnitude of costs had not been developed, in part, because certain data are missing, such as the number of mechanical pumps still in use nationwide. Estimates of the cost to inspect the new equipment as part of state enforcement of weights and measures standards varied. Officials in a small number of states said inspection times would increase by 20 to 50 percent, while in three other states, officials said the costs would not be significant. However, none of these officials had estimated the costs. Finally, although adopting temperature compensation would require that consumers be educated about it, no stakeholders had developed estimates of the costs to, for example, provide disclosure on street signs, fuel pumps, and customer receipts.

Stakeholders differ on whether retailers, consumers, or both would ultimately end up paying the implementation costs. For example, some stakeholders, including state officials and industry representatives, said that the costs of implementing automatic temperature compensation would be passed on to consumers. In their view, the costs to purchase and install compensation equipment would be passed on to consumers through higher prices for fuel or other goods purchased at retail fueling stations. Other stakeholders, such as consumer groups, said that retailers and consumers would share both the costs and the benefits of implementing temperature compensation. That is, one stakeholder said some retailers could use funds provided to them by major oil companies for remodeling to pay for the equipment. Consumers, they say, currently pay retailers for energy content they do not receive when they buy fuel that is warmer than 60 degrees F. Moreover, these stakeholders said that consumers would gain by receiving more consistent energy content, and one said that retailers would benefit because the automatic temperature compensation technology would make it easier to detect gas leaks and to manage inventory. Stakeholder views were based on professional judgment, general economic theory, and assumptions about how the fuel market operates rather than on studies or other data, and most stakeholders said that a comprehensive cost-benefit analysis would provide policymakers with important information.

Governments that have adopted or allowed automatic temperature compensation cited improved measurement accuracy and greater equity between retailers and consumers as reasons for making the change, whereas those that had not adopted automatic temperature compensation cited concerns that the costs would outweigh the benefits. For example, Hawaii adopted temperature compensation more than 26 years ago because, according to Hawaiian officials, it provided purchasing equity for both the industry and the consumer. According to Belgian officials, Belgium mandated temperature compensation beginning in January 2008 to help ensure greater consistency in the energy content of the fuel sold to consumers. To improve measurement accuracy and equity, among other things, Canada developed standards in the early 1990s that allowed, but did not require, retailers to sell temperature-compensated fuel, according to a Canadian official. In the United States, officials from eight states that prohibited automatic temperature compensation said the decision should be based on an analysis of the costs and benefits, with some expressing concern that the anticipated costs would outweigh any benefit to consumers and fuel retailers. Governments have not formally studied the impact of their decisions to implement or allow automatic temperature compensation. Specifically, neither Hawaii nor Canada has studied the impact of temperature compensation, although officials reported it had been well accepted by both consumers and the industry and was not controversial. In Belgium, temperature compensation has not been in effect long enough to study its impact.

BACKGROUND

From the beginning of the modern petroleum industry in the early 1900s, both industry and the federal government have recognized the problem that temperature-induced changes in volume present for inventory control. Specifically, the fact that petroleum products, like most other substances, expand when heated and contract when cooled means that the amount of fuel in the inventories of retailers changes, literally, with the weather. Following a study of the issue conducted by the American Petroleum Institute from 1912 to 1917, the United States and Great Britain established the standard measure for petroleum products: at an ambient temperature of 60 degrees F, 231 cubic inches equals a gallon.

The effect of temperature on fuel volume varies depending on the density of the fuel. For example, gasoline's volume changes approximately 1 percent for every 15 degree temperature change, whereas diesel, which is a more dense fuel, changes approximately 1 degree temperature change. In practice, the density of gasoline and diesel sold to consumers varies depending on such things as the crude oil used to produce the fuel and the addition of other components to achieve certain ends. For example, federal efforts to reduce petroleum consumption and greenhouse gas emissions require the increased use of some components in fuel blends, such as ethanol, biodiesel, and other alternative fuels. In addition, ethanol is added to gasoline in certain geographic areas to help reduce the emissions that contribute to the formation of ground-level ozone, which has been linked to respiratory and other health problems. As a result, the composition and density of gasoline and diesel products vary considerably across the country. In 2004, at least 45 different kinds of gasoline were produced in the United States.

Certain properties of fuels other than volume, such as mass and energy content, do not change in response to changes in temperature. However, energy content can be affected by changes in the density of fuel that arise from the addition of alternative fuels or other blending components that have densities different from the gasoline itself.

In the United States, the petroleum industry often adjusts for temperature-related changes in wholesale transactions for gasoline and diesel and in retail sales for other petroleum products, such as home heating oil, liquefied petroleum gas, and prepackaged liquids such as motor oil. In contrast, virtually all gasoline and diesel sold at the retail level is sold at 231 cubic inches per gallon regardless of the temperature of the fuel.

Temperature compensation can be achieved through several methods. First, volumetric changes can be calculated manually when the fuel density and temperature are known. Second, technological advances have led to the development of devices that automatically measure both the volume and temperature of the fuel at the time of purchase and correct the volume to the amount that would exist if the fuel were at 60 degrees F. Finally, in areas where the ambient temperature remains relatively constant throughout the year, pumps can be recalibrated to dispense the volume a gallon would occupy at 60 degrees F. For example, if the temperature in an area is relatively constant at 75 degrees F, pumps can be recalibrated to dispense 233 cubic inches per gallon.

Gasoline and diesel are distributed nationwide to fuel wholesalers through a supply infrastructure composed of pipelines, barges, tanker vessels, marine terminals, railroads, trucks, and storage tanks. At various points along the distribution chain, fuel is stored at terminal stations that generally have several large storage tanks. Fuel is then distributed, usually by trucks, to retail gasoline stations, where it is typically stored in underground tanks (see Figure 1).

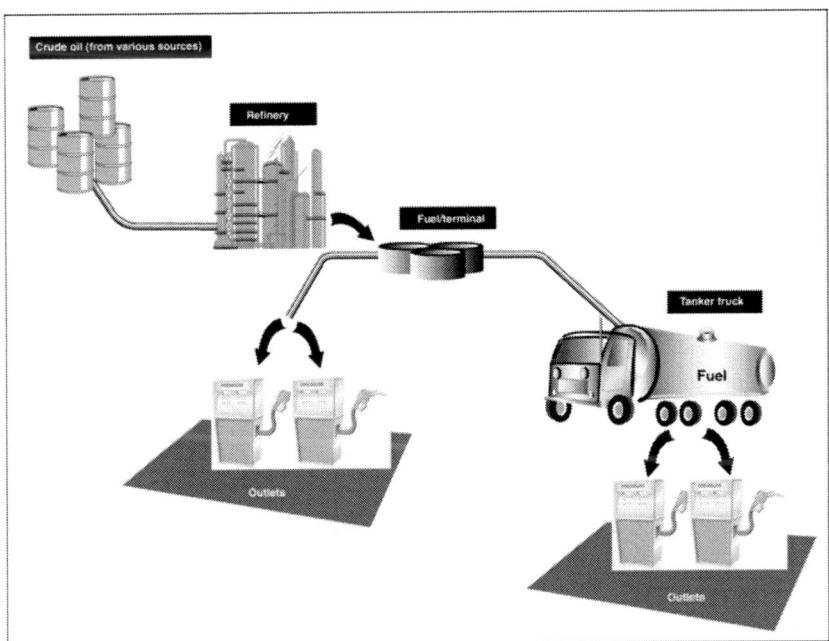

Sources: GAO and Art Explosion (clip art).

Figure 1. Distribution Network for Gasoline and Other Petroleum Products.

Changes in the temperature of gasoline and other petroleum products can occur for several reasons from the time these products leave the refinery until they are deposited into a vehicle. For example, retail fueling stations located near a refinery or a pipeline may receive fuel that is still hot from the refining process, and the heated fuel will affect the temperature of the fuel already in the storage tank.[5] In addition, the use of underground storage tanks—particularly those with double walls—may lengthen the time required for the fuel to cool to ground temperature of about 55 degrees F. A common misconception is that the use of underground storage tanks helps ensure that fuel remains at or below 60 degrees F. According to a 2004 NIST study based on 2 years of data, the average temperature nationwide for fuel stored underground was about 64 degrees and varied among states from about 82 degrees in Florida to 53 degrees in Minnesota. Finally, the temperature of the fuel in the supply line to the pump will affect the temperature of the fuel initially deposited into the vehicle.

State and local governments adopt and enforce weights and measures regulations, including those to ensure that retail fuel pumps accurately measure motor fuels. Unlike many other countries, the United States does not have a federal weights and measures regulatory agency, although two federal agencies help oversee the marketplace generally, and a third oversees aspects of the retail petroleum industry. Among other things, NIST cooperates with other entities, including state and local governments, to establish standard practices, codes, and specifications. The FTC enforces consumer protection laws, including laws related to unfair and deceptive practices in the marketplace. EPA and authorized states regulate underground storage tanks that store petroleum.[6] These regulations require a leak detection system on the underground storage tanks. None of these agencies has formally endorsed or opposed the implementation of automatic temperature compensation.

State and local governments develop regulations for weights and measures with input from NCWM and NIST. Established in 1905, NCWM is composed of state and local weights and measures officials, as well as related public and private sector members. A key goal of NCWM is to help ensure that consumers get the quantity of goods they pay for and that businesses sell the quantity that they advertise and intend to sell. NCWM helps ensure that uniform standards are applied to commercial transactions by developing regulatory standards for consideration by each jurisdiction, with technical, scientific, and administrative support provided by NIST. Membership in NCWM is open to all interested individuals, including regulatory officials, device manufacturers, and consumers; however, only regulatory officials may vote on the disposition of proposals under consideration by NCWM.

Most proposals for regulatory standards that come before NCWM originate in one of its regional weights and measures groups located throughout the nation or in one of its four standing committees, each of which focuses on a specialized area, such as laws and regulations. At each of NCWM's annual conferences, standing committees review the proposals submitted for consideration and hold open hearings to discuss them. Final reports containing the NCWM-approved model regulatory standards are presented in open forum to representatives and voted upon. Actions or subjects under consideration, but not proposed for voting, may be carried over for further consideration at a later time. NIST publishes NCWM's newly adopted model regulatory standards in handbooks. If a state chooses to adopt the model regulatory standards in state law or regulation, they will then have the effect of law in that state.

For over 30 years, NCWM has debated the pros and cons of compensating for temperature-induced changes in the volume of petroleum products, including gasoline and diesel. This debate is guided in part by NCWM's principles that any method of sale or measurement must provide accurate and adequate information about products so that purchasers can make price and quantity comparisons. In 2007, a standing committee recommended a proposal to allow, but not require, automatic temperature compensation at the retail level. NCWM did not reach consensus on the proposal, and the issue was deferred for further consideration. In 2008, a steering committee established by NCWM recommended a proposal to require automatic temperature compensation following a 10-year period during which retailers could decide when to purchase the equipment based on their business needs. According to the committee, this would allow the marketplace to determine when and whether to adjust retail sales for temperature. However, NCWM members did not reach consensus on the proposal, and the issue was deferred for further consideration. Also in 2007, the California legislature directed the state Energy Commission to study the costs and benefits of using automatic temperature compensation devices for retail sales, among other things. The commission is to complete its work by February 2009.

THE MAGNITUDE OF EQUIPMENT AND EDUCATION COSTS OF ADOPTING AUTOMATIC TEMPERATURE COMPENSATION IS UNCLEAR

Stakeholders said that implementing automatic temperature compensation for retail fuel sales would involve costs to purchase, install, and inspect new equipment on gasoline pumps, as well as costs to educate consumers about the change. Some stakeholders estimate the costs to purchase and install the temperature compensation devices would range from $1,300 to $3,000 per pump. To provide context for the estimates from stakeholders, we obtained information from two equipment manufacturers. These manufacturers said the costs can vary by the type of equipment. More specifically, the price of retrofit kits for electronic pumps ranges from

$900 to $1,800, plus $200 to install them. Costs to retrofit mechanical pumps are higher: $2,000 to purchase and install a kit for one hose and $3,800 for a dual hose pump. The costs to individual retailers would vary, in part, depending on the number of pumps, the number of hoses per pump, and the mix of electronic and mechanical pumps that would need to be replaced or retrofitted. In addition, an equipment manufacturer said that maintenance costs for electronic pumps would be negligible over the useful life of a pump, 10 to 12 years. Some stakeholders noted that the magnitude of costs has not been estimated, in part, because certain data, such as the number of mechanical pumps still in use across the country, are not available. As a result, the costs to adopt automatic temperature compensation are not known.

Several stakeholders said costs to purchase and install temperature compensation equipment could also be affected by other factors. For example, under a phased implementation schedule, retailers could upgrade their equipment in the normal course of replacing equipment, whereas immediate implementation would require retailers to invest in the equipment without regard to their business plans or ability to pay immediately. Also, a small number of companies in North America manufacture new pumps equipped to automatically compensate for temperature or kits to retrofit existing pumps. Two stakeholders said that the costs to purchase and install the equipment could rise in the face of shortages of both equipment and skilled installers that would occur if implementation of automatic temperature compensation were to occur suddenly rather than over a longer period of time.

Estimates of the magnitude of inspection costs varied. A small number of state officials said that automatic temperature compensation could increase inspection time by 20 to 50 percent and might require the purchase of testing equipment. In contrast, officials in three other states said that inspection costs to adopt temperature compensation would not be significant, although they had not estimated the cost. In Missouri, state officials said legislation was introduced, but not enacted, to divide the state into regions, each of which would adopt a new reference temperature based on its average ambient temperature. State officials reported that adoption of temperature compensation by changing reference temperatures would require increasing staff by six inspectors and one clerical person for a cost of about $1 million in the first 3 years.

No stakeholders have developed estimates of the costs to educate consumers when automatic temperature compensation is in use. However, costs would be incurred to provide disclosure on fuel pumps, customer receipts, and the street signs that show the retail price of fuel. A number of stakeholders noted that, if some retailers sold compensated fuels and others did not, consumers could be confused and might lack the ability to make informed value comparisons for their fuel purchases. According to some stakeholders, disclosure on pumps might be accomplished by adding the phrase "Volume corrected to 60 degrees F" to the face of the pump near the display of total gallons purchased. For customer receipts, printers could be programmed to add the same phrase. If automatic temperature compensation is in place throughout the nation, the need to disclose its use on pump signs might no longer be needed.

IT IS UNCLEAR WHO WOULD BEAR THE COSTS OF IMPLEMENTING AUTOMATIC TEMPERATURE COMPENSATION

Stakeholders differ on whether consumers or a combination of retailers and consumers would bear the costs of implementing automatic temperature compensation. Specifically, many stakeholders, including state officials and industry representatives, said that the costs to purchase, install, and inspect compensation equipment would be passed on to consumers, generally through higher retail fuel prices, higher prices for nonfuel goods sold at retail fueling stations, or a combination of both. A few of these stakeholders said that retail prices must generally reflect the cost of goods sold or businesses will not remain in operation. However, since the information retailers use to make pricing decisions is proprietary in nature, it would be difficult to estimate how much prices would increase to cover the costs of implementing automatic temperature compensation. Some of these stakeholders also noted that differences in the cost of fuel and other goods sold could vary among retailers based on such factors as whether they owned or leased the land, the number of staff they employ, and whether the costs of inspections are paid directly by retailers or funded from tax receipts. However, one state official said that the ability of states to increase inspection fees may be limited by state statute.

Some stakeholders said the costs to implement automatic temperature compensation may result in disproportionate economic impacts on certain classes of retailers, such as small retailers and those in rural areas, that might be put out of business in the face of the investment to upgrade their equipment. Retailers that are small or located in rural areas may dispense fewer gallons of fuel than larger retailers and, consequently, have fewer gallons from which to recover any costs associated with upgrading their equipment. A few stakeholders said an exemption for small retailers may be needed, such as an exemption based on the number of gallons dispensed. In contrast, another stakeholder said implementation that allowed retailers to make the decision of whether to add the devices to their equipment would eliminate the potential for disproportionate impacts.

However, other stakeholders, such as consumer groups, said that retailers and consumers would share in both the costs and the benefits of implementing temperature compensation. For example, one stakeholder noted that some retailers could use funds they receive from the major oil companies for remodeling to cover the cost of temperature compensation equipment. According to these stakeholders, consumers have already paid retailers for energy content they did not receive. That is, consumers generally buy fuel that is warmer than 60 degrees and has less energy content, according to these stakeholders. Such overpayments are greater in southern and western states than in other areas. Moreover, these stakeholders said consumers would benefit from greater transparency in fuel pricing, the ability to purchase fuel with more consistent energy content, and an enhanced ability to compare purchases from competing retailers because price differences would be based largely on differences in customer service or amenities such as clean rest rooms. One noted that retailers would also benefit because the automatic temperature compensation technology would allow retailers to manage inventory for both their deliveries and their sales of fuel on a temperature-compensated basis. Moreover, retailers could more easily identify fuel leaks by reconciling their inventory records to measurements of the fuel in their storage tanks. Specifically, if a measurement of stored fuel showed a retailer had less fuel on hand than it had sold, the difference could be the result of a leak.

Stakeholders also differed on the benefits of automatic temperature compensation. Many noted that temperature compensation provides a more accurate and replicable measurement method, but some expressed concern that the potential cost outweighed the benefit. Within the weights and measures community, support has been growing for the adoption of automatic temperature compensation standards, in part because of the improved accuracy and the availability of equipment that makes implementation more feasible than in the past. Several stakeholders noted that automatic temperature compensation brings equity to the marketplace and provides both consumers and retailers with comparable information about their fuel purchases. Specifically, when retailers and consumers purchase temperature-compensated fuel, they each receive comparable products. According to two stakeholders, consumers currently cannot determine before or after a purchase the actual best price for a gallon of gas because they do not know the temperature of the fuel. Some stakeholders who thought the cost would outweigh the benefit said that the increased accuracy in measurement would not benefit consumers because fuel costs would increase as retailers recouped their investment in the compensation devices.

Stakeholders also held different opinions on whether automatic temperature compensation would ensure consistent energy content in each gallon of fuel. While temperature compensation adjusts for the impact of fuel temperature on the energy content of each gallon, it would not affect other factors that impact energy content, such as the use of fuel blends and additives. That is, multiple stakeholders said that the use of ethanol and other additives, as well as seasonal fuel blends, results in fuels that may vary in energy content by season or by retail outlet. More specifically, they noted other factors may affect the energy content of fuel, including the refining process itself and the crude oil used as the source for the gasoline. Others said automatic temperature compensation will ensure greater consistency in energy content and mileage per gallon. One stakeholder said that, as fuel prices increase, the issue of energy loss from the lack of temperature compensation will become more important, while another said that the use of blends could increase the significance of the effect of temperature on fuel in the future.

Stakeholders' views that various factors may affect fuel prices are consistent with our prior work on gasoline pricing.[7] Specifically, in a series of reports issued from 2000 through 2007, we concluded that higher gasoline prices resulted from a range of local and global factors, including higher crude oil prices, recent mergers and increased market concentration in the petroleum industry, the increased use of blended fuels, the level of state gasoline taxes, and costs to transport gasoline from refineries to retailers. We also found in our work on the use of special gasoline blends that it can be difficult to establish a definitive causal link between factors and prices because only some of the many factors that may affect gasoline prices at various times are readily and consistently observable.

Regardless of their views, stakeholders based their opinions largely on professional judgment and general economic theory or assumptions about how the fuel market operates rather than on studies or other data. For example, one stakeholder commented that it was unreasonable to assume that retailers would absorb the costs to upgrade 14 or 16 pumps without trying to recoup those costs through the prices of retail goods they sell. However, none of the stakeholders based their views on studies of the impact of the costs on fuel or retail goods. Some stakeholders said that because the petroleum market is fiercely competitive, particularly in areas that sell high volumes of fuel, consumers already receive the lowest fuel price that retailers can offer, and one said that temperature is not likely to be a relevant factor in their pricing decisions. Because the fuel market is so competitive, one stakeholder said, retailers do not generate enough profit to cover the costs of temperature compensation equipment and so would pass the costs on to consumers. In contrast, other stakeholders said that retailers may already adjust their prices to account for the expansion and contraction of fuel, while still others questioned the benefit to consumers from investing in temperature-compensating devices in areas where the average ambient temperature is close to 60 degrees F.

The majority of stakeholders—including state officials, consumer and industry representatives, and fleet owners—said that a cost-benefit study such as the one under way in California would provide policymakers with important information. The California study will examine the costs for retailers to purchase and install appropriate equipment and calibrate it. In addition, the study will develop data on the costs to agencies to develop appropriate test procedures, acquire calibration equipment, and inspect the pumps at retail stations. Information on the costs and benefits was needed to make an informed decision on automatic temperature compensation, according to many stakeholders. A small number said they would wait to see the results of California's study before deciding whether to support or oppose the implementation of automatic temperature compensation. Moreover, some who oppose automatic temperature compensation said they would support it if a cost-benefit analysis showed a benefit for the consumer.

GOVERNMENTS THAT HAVE ADOPTED AUTOMATIC TEMPERATURE COMPENSATION DID SO LARGELY TO IMPROVE PURCHASING EQUITY, AND THOSE THAT HAVE NOT CITED CONCERNS THAT THE COSTS WOULD OUTWEIGH THE BENEFITS

Governments that have adopted or permitted automatic temperature compensation, or are considering doing so, cited improved measurement accuracy and greater equity between retailers and consumers as reasons for making the change, whereas those governments that do not allow temperature compensation cited concerns that the costs would outweigh the benefits. Hawaii, Belgium, Canada, and the European Union (EU) have each adopted a policy on temperature compensation—mandatory in Hawaii and Belgium and permissive in the remaining jurisdictions. In addition, the United Kingdom is considering a national approach to temperature compensation, and Australia may do so again. Both countries debated the issue in the 1990s but did not adopt nationwide policies for retail fuel sales at that time.

Because retail motor fuel dispensers equipped with automatic temperature compensation devices were not readily available 26 years ago, Hawaii developed its own method to achieve temperature compensation for retail sales of fuel to provide purchasing equity for both the industry and the consumer, according to a state official. The method is based on test procedures that rely on both the temperature and density of the fuel. A 5-year study of the average temperature of fuel delivered to consumers in Hawaii found that the fuel temperature was approximately 80 degrees F. More specifically, Hawaiian weights and measures officials test retail pumps to ensure that they meet the state standard—to deliver the amount of fuel a 231 cubic inch gallon would occupy at 60 degrees F, or its expanded or contracted equivalent at any other temperature. In Hawaii, the expanded equivalent is about 234 cubic inches per gallon—to reflect the increased volume at the higher fuel temperature. Implementation was phased in over 1 year. A state official said retailers may apply for a variance from the state standard provided they can demonstrate that the temperature of the fuel they deliver to consumers in their location differs from 80 degrees F. According to a state official, temperature compensation is a matter of fairness and equity.

Belgium mandated temperature compensation for retail sales of fuel beginning in January 2008. Belgium adopted temperature compensation for retail sales, in part, because some retailers were artificially heating fuel, and the government sought greater consistency in the energy content of the fuel sold to consumers, according to a weights and measures official. After January 2008, any newly installed pumps must be equipped for temperature compensation and, by January 2015, all pumps must be equipped to dispense temperature-compensated fuel. A Belgian official told us that the 7-year transition period will allow retailers to make adjustments over time, in the normal course of their operations, thereby reducing the overall cost to implement temperature compensation. While retailers will decide when to install temperature compensation equipment, they are prohibited from turning it off. That is, once the equipment is in place and dispensing temperature-compensated fuel, all hoses attached to the equipment must continue to dispense temperature-compensated fuel. To date, the Belgian government has not developed guidance for consumers or retailers, in part because the transition to temperature compensation has just begun, according to the official.

Canada has adopted a permissive policy on automatic temperature compensation for the retail sale of liquid petroleum products, such as gasoline, diesel, and home heating oil. Specifically, Canada established technical and other standards in the early 1990s that allowed retailers to sell temperature-compensated fuel, but it did not require them to do so. According to a Canadian official, Canada developed the standards largely for three reasons: advances in measurement technology had made temperature compensation equipment more readily available, automatic temperature compensation is thought to be a more equitable and accurate method of measuring fuel, and temperature compensation addresses retailers' concerns about inventory losses potentially due to temperature-related changes in volume. Today, over 90 percent of Canadian fuel retailers sell temperature-compensated fuel. Canada imposed policy controls on the use of temperature-compensated equipment to prevent practices that might harm consumers or businesses, and any change to pumps requires an inspection by government officials. For example, pumps with automatic temperature compensation devices must be operable and dispense temperature-compensated fuel at all times throughout the year. In addition, pumps equipped with the devices must have a sticker that says "Volume Corrected to 15 degrees C"[8] adjacent to the pump's visual and printed net quantity display. Retailers may elect to convert only selected pumps or product lines, provided that all pumps for the same grade or blend of fuel are converted and the compensation equipment is activated at the same time.[9] Because Canada's regulations are permissive rather than mandatory, retailers may choose to stop using compensation devices provided they obtain permission from Canadian weights and measures officials. Permission would not be granted if retailers wanted to only use automatic temperature compensation seasonally when it was to their benefit, according to a Canadian official, who also said no retailers have sought to stop using the devices.

In addition, the EU currently permits member states to adopt temperature compensation, although fewer than 2 percent of retailers have installed the necessary equipment, according to an official with a European weights and measurement organization. This official said that making adoption possible, but not required, allows the market to make the decision when business owners decide it is in their interests to do so. As a result, implementation will occur gradually, thereby avoiding a "shock wave" from immediate mandatory implementation, according to the official. A shock wave would occur if retailers were required to purchase the equipment without regard to whether they had the funds to do so. The EU does not have a harmonized policy on temperature compensation, but, according to the official we interviewed, information on fuel temperature received by the retailer and dispensed to consumers would be important to the debate. However, the official also noted that retailers may, at their discretion, adjust prices to compensate for temperature-related changes in volume.

Currently, in Australia the states and territories require retailers to sell fuel on a compensated basis. However, by July 2010, responsibility for weights and measures regulation will shift from the states and territories to the federal government. According to an Australian official, the new national trade measurement legislation will replicate the current state and territory requirements for the sale of fuel. As part of the consultation process for developing new trade measurement regulations, comments on any aspect of trade measurement controls, such as temperature compensation, will be invited from all stakeholders, and the matter of temperature conversion of fuel sales at the retail level may well be raised.

Officials in the United Kingdom said they anticipate issuing a statement in the fall of 2008 that temperature compensation for retail fuel sales will be permitted nationwide but not mandated.

In the United States, officials in eight states that have laws or regulations prohibiting automatic temperature compensation largely said the decision should be based on an analysis of the costs and benefits, with some expressing concern that the anticipated costs would outweigh any benefit to consumers and fuel retailers. In some cases, these decisions were made more than 20 years ago, and the officials we interviewed had limited information about the reasons. More recently, Missouri and Texas considered state legislation to implement temperature compensation. In Missouri, where the average temperature of stored fuel is 62 degrees F, officials said that consumers would be negatively affected if temperature compensation were adopted by changing the reference temperature because they would have to buy more temperature-adjusted gallons than uncompensated gallons to obtain the same amount of fuel. In addition, the state would need to add six inspectors and one clerical person at a cost of about $1 million in the first 3 years. Moreover, they said retailers would face significant expense to purchase the compensation equipment if temperature compensation were achieved by the use of compensation devices. Specifically, Missouri officials in 2006 estimated that 65 percent of the state's pumps could be retrofitted, and 35 percent would need to be replaced, at a cost of about $341 million. In Texas, officials have postponed further consideration of temperature compensation until a comprehensive nationwide cost-benefit analysis has been completed. In addition, officials in some states said that evidence of benefits to consumers from automatic temperature compensation could lead states to reconsider their current position.

Finally, governments have not formally studied the impact of their decisions to implement or not to implement automatic temperature compensation. Specifically, neither Hawaii nor Canada has studied the impact of temperature compensation, although officials reported it was not controversial and was generally well accepted by both consumers and the industry. In Belgium, temperature compensation has been implemented too recently to study its effects.

CONCLUDING OBSERVATIONS

The weights and measures community has debated the costs and benefits of automatic temperature compensation for more than three decades with no resolution. The issues have not changed substantively, and both sides continue to passionately put forth their views. In general, supporters say that extending temperature compensation to the retail level could provide more transparency in fuel prices, while those who are opposed argue that upgrading existing equipment would be costly and pose potential economic hardship on retailers.

It remains unclear, however, what it would actually cost to implement automatic temperature compensation and whether consumers or businesses would end up paying those costs. Moreover, the two governments with the longest experience in temperature compensation of retail fuel sales (Hawaii and Canada) have not studied the effect of their policies. As a result, a policy debate is being played out without good information about the potential costs and benefits, and with both proponents and opponents basing their views on their professional judgment and their general understanding of economic theory.

Looking forward, there appears to be a real need for an objective analysis of the key issues stakeholders raise about costs and benefits, including the potential for higher costs to consumers and improved inventory management for retailers. Such a study would need to bring together petroleum-related scientific, engineering, and economic expertise. Absent such analyses, NCWM and state governments face potentially significant challenges to informing their decisions regarding automatic temperature compensation.

As agreed with your office, unless you publicly announce the contents of this report earlier, we plan no further distribution until 30 days from the report date. At that time, we will send copies of this report to the Chief, Weights and Measures Division, National Institute of Standards and Technology; stakeholders we interviewed; appropriate congressional committees; and other interested parties. We will also make copies available to others upon request. In addition, the report will be available at no charge on the GAO Web site at http://www.gao.gov.

If you or your staff have any questions about this report, please contact me at (202) 512-3841 or maurerd@gao.gov. Contact points for our Offices of Congressional Relations and Public Affairs may be found on the last page of this report. GAO staff who made contributions to this report are listed in appendix II.

Sincerely yours,

David C. Maurer
Acting Director
Natural Resources and Environment

APPENDIX I: SCOPE AND METHODOLOGY

In conducting our work on each of our objectives, we reviewed National Conference on Weights and Measures (NCWM) documents and congressional testimony and performed a literature search to identify relevant documents and stakeholders likely to have a view on the implementation of automatic temperature compensation. We used the individuals identified through our document review and literature search as a starting point for the sampling technique that we used to identify additional stakeholders. That is, we used an iterative process (often referred to as the "snowball sampling" technique) to identify other stakeholders and selected for interviews those who would provide us with a broad and balanced range of perspectives on temperature compensation of gasoline and diesel. We used a standard set of questions to interview each of these individuals to ensure we consistently discussed each aspect of automatic temperature compensation. We also asked open-ended questions to allow people to share their views on this issue. To develop the questions, we reviewed NCWM and National Institute of Standards and Technology (NIST) documents, as well as congressional testimony. We used content analysis to identify the main themes among responses.

We continued interviewing and soliciting names until we determined we had appropriate coverage from all the relevant stakeholder groups. During the course of our review, we interviewed officials from the following 15 organizations, listed alphabetically: American Automobile Association; American Petroleum Institute; American Trucking Association; Consumer Watchdog; Defense Energy Support Center; National Association of Convenience Store Owners; NATSO, an organization representing travel plaza and truck stop owners; Owner Operator Independent Drivers Association; Petroleum Marketing Association of America; Society of Independent Gasoline Marketers of America; Schneider National, Incorporated; Swift Transportation Incorporated; United Parcel Service; United States Postal Service; and Weights and Measures Consulting. In addition, we interviewed federal officials at NIST, the Environmental Protection Agency, and the Federal Trade Commission because these agencies help oversee the marketplace generally or oversee aspects of the retail petroleum industry. We also obtained information from two of the three manufacturers who produce equipment that allow for automatic temperature compensation at retail pumps.

We also contacted officials in 17 states that the literature suggested may have taken or considered specific steps to adopt or prohibit automatic temperature compensation. Some of these states had proposed legislation, were identified in a survey conducted by NIST on state practices, or were recommended by other officials. One state—California—is conducting a state-mandated cost-benefit analysis of automatic temperature compensation. These 17 states included a mix of states that could be considered hot (5), cold (4), or neutral (7) based on NIST's analysis of temperature data for stored fuel. The 17th state was not included in NIST's analysis because of a lack of data. We interviewed officials in the following 17 states, listed alphabetically: Arizona, California, Florida, Hawaii, Iowa, Maine, Massachusetts, Minnesota, Missouri, Montana, Nebraska, New York, Oregon, Pennsylvania, South Dakota, Texas, and Wyoming.

Finally, we interviewed officials in Australia, Belgium, Canada, the Netherlands, and the United Kingdom because literature indicated they either had adopted or had considered implementing automatic temperature compen-sation, as well as officials at a European weights and measures organization.

We conducted our work from March 2008 to September 2008 in accordance with generally accepted government auditing standards. Those standards require that we plan and perform the audit to obtain sufficient, appropriate evidence to provide a reasonable basis for our findings and conclusions based on our audit objectives. We believe that the evidence obtained provides a reasonable basis for the information we present for each of our audit objectives.

ENDNOTES

[1] This report focuses on gasoline and diesel rather than other petroleum products, such as heating oil or jet fuel.

[2] This example assumes the use of the same blend of gasoline. Energy content can also vary depending on the blend of gasoline.

[3] Throughout this report, we refer to the devices that dispense fuel as pumps. Individual pumps may dispense multiple types of fuel, such as regular and high-octane gasoline.

[4] Throughout this report, we use the word "stakeholder" to refer to domestic individuals and groups with an interest in the current debate in the United States on this issue, including NCWM, NIST, current and former government officials, consumer groups, representatives of the petroleum and trucking industries, and fuel retailers.

[5] The refining process "boils" crude oil to separate it into its various components. Gasoline is distilled from crude oil at temperatures ranging from 194 degrees F to 428 degrees F, while diesel is distilled at temperatures up to 698 degrees F.

[6] The underground storage tank regulations apply to underground tanks and pipes used to store or move petroleum and certain other hazardous chemicals.

[7] GAO, *Energy Markets: Increasing Globalization of Petroleum Products Markets, Tightening Refining Demand and Supply Balance, and Other Trends Have Implications for U.S. Energy Supply, Prices, and Price Volatility,* GAO-08-14 (Washington, D.C.: Dec. 20, 2007); GAO, *Gasoline Markets: Special Gasoline Blends Reduce Emissions and Improve Air Quality, but Complicate Supply and Contribute to Higher Prices,* GAO-05-421 (Washington, D.C.: June 17, 2005); GAO, *Energy Markets: Mergers and Many Other Factors Affect U.S. Gasoline Markets,*

GAO-04-951T (Washington, D.C.: July 7, 2004); GAO, *Motor Fuels: Gasoline Prices in Oregon,* GAO-01-433R (Washington, D.C.: Feb. 23, 2001); and GAO, *Motor Fuels: California Gasoline Price Behavior,* GAO/RCED-00-121 (Washington, D.C.: Apr. 28, 2000).

[8] The reference standard of 15 degrees Celsius (C) is roughly equivalent to 60 degrees F.

[9] Canada also allows partial conversion to automatic temperature compensation based on "trade levels" that use different types of pumps, such as those mounted on vehicles or those that dispense fuel at high speed. In such cases, all pumps for a given trade level must be converted and activated at the same time.

In: Fuel Prices: Rhyme or Reason?
Editor: William P. Vestus

ISBN: 978-1-60692-842-4
© 2009 Nova Science Publishers, Inc.

Chapter 5

GASOLINE PRICES: CAUSES OF INCREASES AND CONGRESSIONAL RESPONSE

Carl E. Behrens and Carol Glover

SUMMARY

The high price of gasoline has been and continues to be a driving factor in consideration of energy policy proposals. Despite passage of the massive Energy Policy Act of 2005 (EPACT 2005, P.L. 109-58), and the Energy Independence and Security Act of 2007 (H.R. 6, P.L. 110-140), numerous other proposed initiatives remain under active consideration in the 1 10th Congress. Measures proposed include repeal of some tax benefits to domestic oil and gas producers contained in EPACT2005, provisions on price gouging, and reform of oil and gas leasing in the Gulf of Mexico.

A large number of factors have combined to put pressure on gasoline prices, including increased world demand for crude oil and limited U.S. refinery capacity to supply gasoline. The war and continued violence in Iraq added uncertainty, and threats of supply disruption have added pressure, particularly to the commodity futures markets. Concern that speculation has added volatility and upward pressure has frequently been cited. In recent months, a decline in the value of the dollar compared to other currencies has increased the dollar price of oil on futures markets.

The gasoline price surge has stimulated much legislative activity, but until recently there has not been the sense of the extreme urgency of previous energy crises. In part, this may be due to the fact that there has been no physical shortage of gasoline or lines at the pump, as there were after the Arab oil embargo in 1973 and the Iranian revolution in 1979. At that time there was expectation that prices were destined to grow ever higher, and many believed that the world's supply of oil was running out. Such views have been less prevalent during the current run-up. But the continued and unrelenting increase in crude oil prices to record levels, even discounting inflation, is leading many to suggest that changing world market conditions may have led to permanent, or at least chronic, shortages of petroleum production capacity. Others continue to expect that growth in demand will moderate, and production will increase to meet demand, as it did following the shortages of the 1970s.

The continuing high prices have led to a further search for legislative remedies. This report, after analyzing factors that have contributed to high gasoline prices, describes the major legislative initiatives and discusses the issues involved.

Most Recent Developments

Gasoline and crude oil prices surged to record levels in May 2008 but as the summer driving season ended they moderated somewhat. (See **Figure 1**.) Consumption of gasoline continued above 9 million barrels per day (mbd), but cumulative consumption for the first 240 days of 2008 was 185,000 barrels per day less than the same period in 2007.

Despite passage in December 2007 of the Energy Independence and Security Act (H.R. 6, P.L. 110-140), the main provisions of which were an increase in the Corporate Average Fuel Economy (CAFE) standards for automobiles and light trucks, and an increase in the requirement for the use of renewable fuels in gasoline, the latest increases have led to urgent discussion of ways to increase supply and ameliorate prices, in Congress, by the Administration, and on the campaign trail. On May 13, 2008, both the House and the Senate passed legislation that would prohibit the federal government from acquiring oil for the Strategic Petroleum Reserve (SPR) during 2008. The President signed the bill and the Department of Energy announced that fill would cease in July.

In: Fuel Prices: Rhyme or Reason?
Editor: William P. Vestus

ISBN: 978-1-60692-842-4
© 2009 Nova Science Publishers, Inc.

Chapter 3

TRANSPORTATION FUEL TAXES: IMPACTS OF A REPEAL OR MORATORIUM

Robert Pirog and John W. Fischer

SUMMARY

Legislation that would repeal or otherwise provide for a summer-long moratorium of federal transportation fuel taxes has been introduced in the 1 10th Congress. Simultaneously, Senators McCain and Clinton are proposing a summer fuel tax collection moratorium as part of their Presidential campaigns. Fuel prices have risen rapidly in 2008 for a variety of reasons. Those seeking to alter federal fuel tax collection are doing so in the belief that a reduction in fuel taxes would give Americans a modest level of economic relief from high pump prices. Current market conditions and the marginal amount of tax relief incorporated in most proposals, however, raise uncertainty as to whether prices to individuals and businesses would fall and whether any price decline would be meaningful to consumers in economic terms. Also of concern is the possible impact of any repeal or moratorium on the overall federal budget deficit.

A reduction in transportation fuel taxes would result in a decrease in spending for Highway Trust Fund-supported federal programs, unless Congress designated alternate sources of funding for these programs. As a result of the structure of the federal programs, the effects of a fuel tax repeal on federal transportation programs would not necessarily be immediate, but depending on the length and scope of the repeal or suspension, they could be substantial.

Legislation that would repeal or otherwise provide for a summer-long moratorium of federal transportation fuel taxes has been introduced in the 110th Congress. Simultaneously, Senators McCain and Clinton are proposing a summer fuel tax collection moratorium as part of their Presidential campaigns. Fuel prices have risen rapidly in 2008 for a variety of reasons. Those seeking to alter federal fuel tax collection are doing so in the belief that a reduction in fuel taxes would give Americans a modest level of economic relief from high pump prices. There is, however, significant opposition to this proposal by those who believe that a fuel tax "holiday" would provide minimal relief to individuals, while potentially adding to the overall federal deficit.

INCREASE IN CRUDE OIL AND REFINED PRODUCT PRICES

Due to the continued tightness in crude oil markets, spot prices of crude oil rose by almost $20 per barrel between the end of December 2007 and the end of April 2008, from about $91 per barrel to about $112 per barrel. Perhaps more publicized, the futures price of one grade of crude oil exceeded $115 per barrel in late April. The average acquisition cost of crude oil to U.S. petroleum refiners (which excludes some transportation costs) increased by approximately 15% over almost the same period, from $85 per barrel at the end of December 2007 to $98 per barrel in March 2008.

Because the cost of crude oil to refiners accounts for a substantial portion of the price of refined petroleum products to users, retail prices of gasoline, diesel fuel, and heating oil have risen. The average U.S. retail price of gasoline of all grades increased from about $3.07 per gallon at the end of December 2007 to $3.51 per gallon by the end of April 2008, an increase of about 14%. The average price of diesel fuel per gallon increased from about $3.34 in December 2007 to $4.08 in late April 2008, or 22%. Jet fuel prices rose from $2.68 per gallon in December 2007 to $3.10 in late April 2008. Home heating oil prices increased by 25% from December 2007, the beginning of the winter heating season, to April 2008, the end of the winter heating season, from $2.60 per gallon to $3.24 per gallon.

PROPOSALS TO OFFSET EFFECTS OF HIGHER CRUDE OIL PRICES

The steep increases in the retail prices of refined petroleum products over a six- month period have prompted some Members of Congress to seek means of countering the consequences of higher crude oil prices and/or reducing the retail prices. Among other policy options, interest has focused on a possible moratorium on the federal excise taxes on transportation fuels, especially gasoline and diesel fuel.

Virtually all transportation fuels are taxed under a complicated structure of excise tax rates and exemptions that vary by transportation mode and fuel type. Gasoline used in highway transportation is taxed at a rate of 18.4 cents per gallon. This is composed of an 18.3-cent-per-gallon Highway Trust Fund rate, the revenues from which are earmarked for the federal Highway Trust Fund, and a 0.1-cent-pergallon rate dedicated to funding the Leaking Underground Storage Tank (LUST) Trust Fund. Diesel fuel used in highway transportation is subject to total federal excise taxes of 24.4 cents per gallon, 24.3 cents of which is earmarked for the Highway Trust Fund and 0.1 cent of which goes to the LUST fund. Every state also has excise taxes on fuels used for highway transportation; these differ widely by state. Some states have employed fuel tax moratoriums in the past, with varying results.[1]

In the current situation, congressional attention has focused on temporary and extended suspension of highway fuel excise taxes. Higher gasoline costs for consumers reduce the amount of disposable income available for other purchases, and may be especially disruptive to consumers during the summer driving season.[2] Higher fuel costs for truckers potentially increase hauling charges, transportation costs, and consumer prices, and may also decrease trucking company profits and/or drivers' income. Fuel costs constitute a significant portion of trucking company operating costs.[3]

Selected Legislative Proposals in the 110th Congress

S. 2890 (Senator McCain, April 17, 2008) would suspend the fuel taxes for the summer driving season (May 26, 2008, to September 1, 2008), and requires that the Highway Trust Fund and LUST accounts be reimbursed by Treasury general funds for revenues not collected from fuel taxes. S. 2971 (Senator Clinton, May 2, 2008) would suspend fuel taxes for the same period. It would also reimburse the Highway Trust Fund and LUST deposits from the Treasury general funds, although it does provide offsetting new revenues from a windfall profits tax on the oil industry. A third Senate bill, S. 2896 (Senator Snowe, April 21, 2008) would partially suspend only the diesel highway fuel excise tax, setting it at 18.3 cents per gallon from passage until December 31, 2008. This bill would also provide for Highway Trust Fund and LUST reimbursement from Treasury general funds.

Two House bills were introduced in the first session of the 1 10th Congress that take a different approach to providing fuel tax relief. H.R. 1569 (Representative McHugh, March 19, 2007) suspends highway fuel taxes whenever the per-gallon price of gasoline exceeds $2.75. H.R. 2448 (Representative Kuhl, May 23, 2007) reduces the highway gasoline excise tax by 10 cents per gallon whenever the price of gasoline exceeds $3.00 per gallon. H.R. 1569 would provide for Highway Trust Fund and LUST reimbursement; H.R. 2448 would not.

IMPACTS ON MARKETS AND PRICES

As indicated, the measures described are motivated by the steep and rapid increases in the retail prices of refined petroleum products, and are intended to reverse those increases, at least to some extent, for a limited period of time. *Under "normal" market conditions and assuming a reasonable degree of competition in oil and petroleum product markets*, the market response to a cut in the excise taxes would be a tendency to reduce user prices by an amount less than or equal to the tax cut.

Economic theory suggests that the key factor in determining the extent of the pass-through to consumers of the reduced tax is the degree of price responsiveness of the supply of petroleum products.[4] If the quantity supplied is extremely responsive to even tiny price changes, the entire tax cut would be passed on to consumers.[5] If, as is more likely in the real world, there is a less sensitive relationship between prices and the supply of refined petroleum products, less of the tax cut would be passed on to consumers. In the case of no sensitivity of petroleum product supply to changes in price, none of the tax cut is likely to be passed on to consumers. In the latter two cases, the taxed entities — refiners, importers, and terminal operators[6] — would view the tax cut as a decrease in the cost of doing business, take advantage of the quantity-constrained nature of supply, and pass forward some, or none, of the tax cut in the form of lower prices.

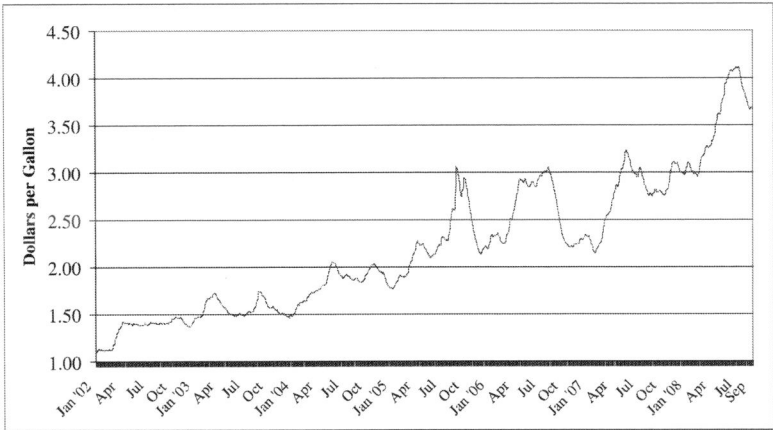

Notes: Prices include federal, state, and local taxes. Last date above is September 5, 2008. Source: *Daily Fuel Gauge Report*, American Automobile Association [http://www.fuelgaugereport.com], compiled by CRS.

Figure 1. Average Daily Nationwide Price of Unleaded Gasoline, January 2002- September 2008.

Energy continued to be an issue in the Presidential campaign. In June Senator McCain reversed his previous position and proposed lifting the moratorium on Outer Continental Shelf (OCS) oil and gas leasing. He continued to oppose leasing in the Arctic National Wildlife Refuge (ANWR). (See below.)

On September 16 the House passed H.R.6899, the Comprehensive American Energy Security and Consumer Protection Act. The bill would replace the moratorium on OCS leasing, which expires October 1, 2008, with provisions allowing limited leasing. However, on September 24 the House passed a continuing appropriations measure that would fund Government activities until March 6, 2009. The measure, H.R. 2638, the Consolidated Security, Disaster Assistance, and Continuing Appropriations Act, 2009, does not include extension of the OCS moratorium.

BACKGROUND AND ANALYSIS

Legislative Activities

The persistence of high gasoline prices led to a broad spectrum of proposed new legislation in the First Session of the 110th Congress. Despite passage of the major Energy Policy Act of 2005 (P.L. 109-5 8), many Members continued to explore a variety of measures to increase supply and reduce demand in the short term, and to reduce the impact of high prices on consumers, as well as revisit longer-term policies that were left behind in the process of reaching agreement on P.L. 109-5 8.

One such proposed policy was increasing CAFE standards for automobiles and light trucks, and the Energy Independence and Security Act of 2007 (H.R. 6, P.L. 110-140) resolved a decades-long debate by setting new standards and procedures for meeting them. P.L. 110-140 also increased the requirement to use renewable fuels in gasoline, including advanced biofuels such as cellulosic alcohol starting in 2016. However, a number of proposals included in one or more versions of energy legislation in 2007 were dropped from the final bill, and those issues remain of interest to the Congress during the Second Session.

With gasoline prices soaring, a new wave of legislative proposals appeared in the Congress. Prominent among them were bills to suspend the federal gasoline and diesel transportation tax during the summer driving season, by presidential candidates Senators McCain and Clinton. Senator Domenici introduced a bill emphasizing U.S. petroleum production, including opening the Outer Continental Shelf (OCS) and part of the Arctic National Wildlife Refuge (ANWR) for oil and gas leasing and encouraging leasing of oil shale deposits. Democrats in both the House and the Senate were reported to be preparing new energy proposals to deal with the situation, and Senator Reid soon introduced a bill which, among other measures, would impose a windfall profits tax on oil companies. Numerous bills have also been introduced to deal with the possibility that speculation is unreasonably driving up oil prices.

This report reviews the major legislative initiatives to deal with the gasoline price issue. To put these proposals in perspective, it first describes some of the factors that have led to the high prices of both crude oil and gasoline.

Why Are Prices So High?

The run-up of gasoline prices that began in spring 2004 (see **Figure 1**) climaxed a period of almost five years during which gasoline prices demonstrated a great deal of regional volatility but less of an increase at the national level. In 2004, a large number of factors combined to exert pressure on gasoline prices in all parts of the country. Some of these factors have affected the price of crude oil, and others the cost of producing and marketing gasoline.

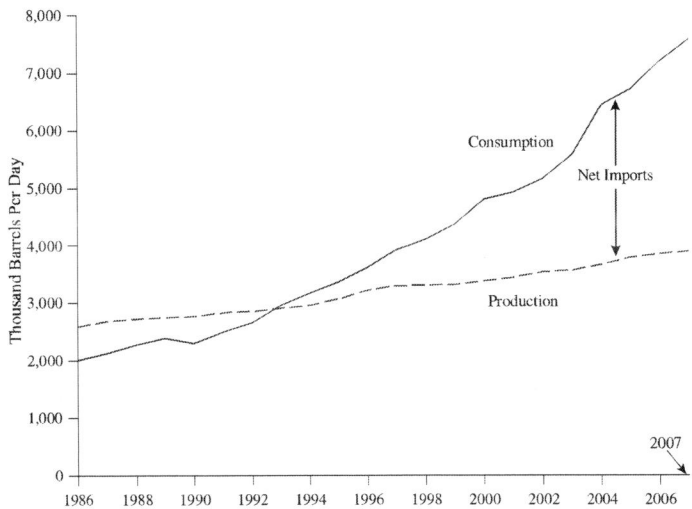

Note: 2007 is the forecast value.
Source: EIA, *China Energy Profile*, downloaded on May 19, 2007. [http ://tonto. eia.doe.gov/country/country _energy _data. cfm?fips=CH].

Figure 2. China's Oil Production and Consumption, 1986-2007.

Crude Oil Prices. Past energy crises have demonstrated that oil is traded in a world market, in which events in remote areas affect the price of crude for almost everyone. As a result, the price of crude oil is set through the interaction of world demand and supply. Major factors in the run-up of crude oil prices have been the sharply increased consumption of imported oil by China (see **Figure 2**) and the continuing possibility of a supply disruption, either from violence or terrorism in the Middle East, or from natural disasters like Hurricanes Katrina and Rita in 2005.

World demand for crude oil grew by 1.3% in 2007 to 86.0 mbd. It is forecast to grow by 1.5% to 87.3 mbd in 2008. World supply was 87.3 mbd in March 2008, leaving relatively little excess supply to draw on if the market were disrupted by natural or political disasters.[1] When excess supply on the market is low, prices tend to rise and become more volatile.

Some observers have suggested that speculators, who have entered the commodity markets in large numbers looking for ways to increase their monetary investments rather than to trade in oil and oil products, are causing an unacceptable upward pressure on prices. Another factor in recent months has been the decline in the value of the dollar compared to other currencies. Since world prices of oil are quoted in dollars, this would have an upward effect on market prices.

One of the major factors pushing crude prices higher is the perception that, as demand increases, production capacity will not increase with it. Most of the spare production capacity in the world market is located in OPEC countries, and, as **Figure 3** shows, spare capacity in those countries has been lower than average over the past several years.

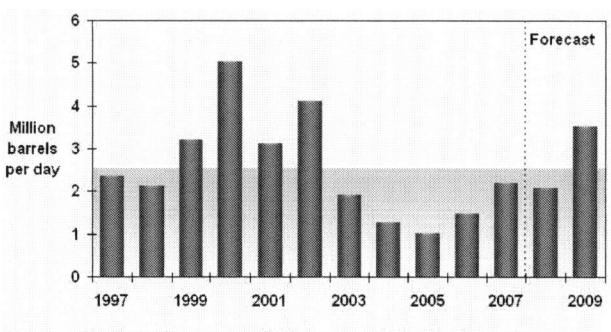

Source: EIA, *Short-Term Energy Outlook*, Figure 10, May 2008.
Figure 3. OPEC Surplus Crude Oil Production Capacity.

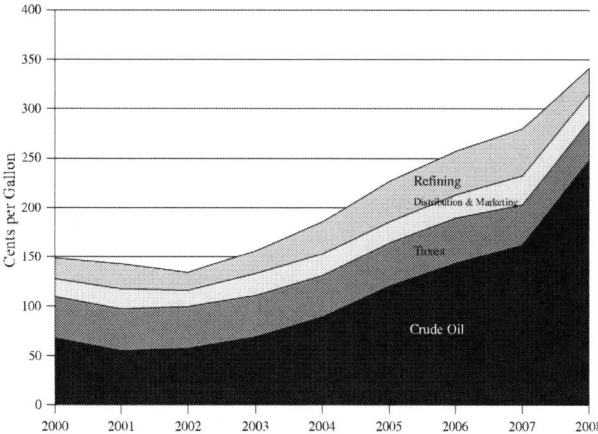

Note: 2008 data is for January through June.
Source: EIA, Gasoline and Diesel Fuel Update, July 30, 2008. Data calculated from monthly percentages by CRS.
Figure 4. Average Annual Components of Gasoline Prices, 2000-2007 and January-June 2008.

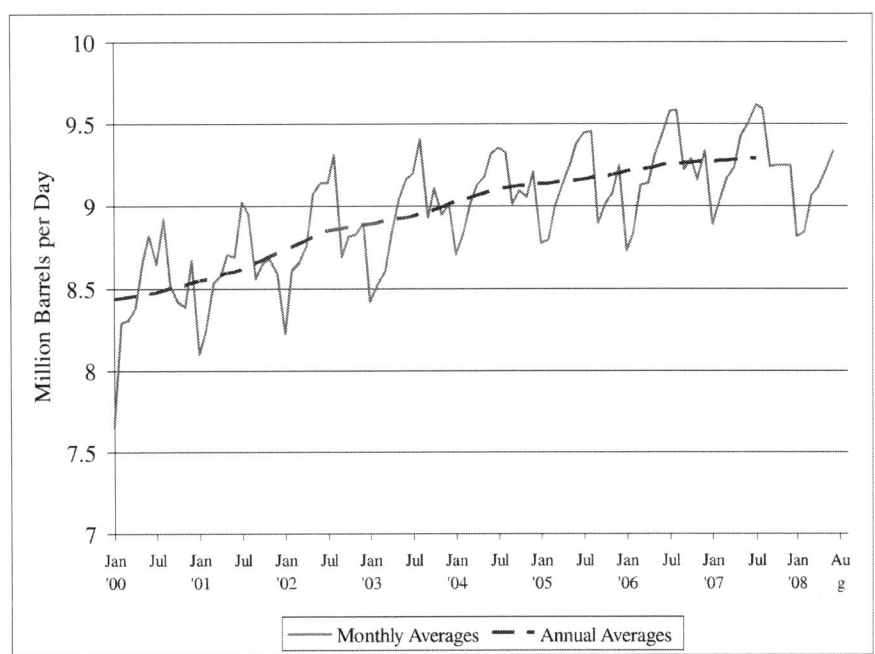

Note: The data point for August 2008 is the average for the four week period ending on 8/29/08.

Source: EIA, *Monthly Energy Review*, August 2008, Table 3.5 and EIA, Weekly Petroleum Status Report, September 3, 2008, Table 10.

Figure 5. U.S. Gasoline Consumption, January 2000 - August 2008.

Gasoline Prices. Higher prices for crude oil tend to translate directly into higher prices for gasoline. Currently, crude oil accounts for about 72% of the cost of gasoline. Refining, distributing, and marketing account for about 16% of the cost of gasoline, and taxes account for about 13%. However, until recently crude oil's share of the cost of gasoline has been more typically in the range of 45% to 55%. In May 2007, for example, with gasoline at $3.15 per gallon, crude oil contributed 46% of the cost; refining, distributing and marketing 41%; and taxes 13%.[2] This trend is illustrated in **Figure 4**.

Whether the crude oil a refiner processes is purchased on the open market or is produced by the oil company itself, higher costs for any element in the cost of gasoline are likely to be passed on to consumers.[3] A number of factors have aggravated the pressure on gasoline prices, including limited refining capacity in the United States, the range of fuel blends required to meet air pollution requirements, and the mandated use of ethanol as an additive. Perhaps most important, U.S. demand for gasoline has increased as economic growth continued, at least through 2007. However, consumption of gasoline for the first 240 days of 2008 averaged 9.12 mbd, compared to 9.30 mbd during the same period in 2007.[4] (See **Figure 5**.)

The 2004 price surge intensified discussion of energy policy and led to further calls for passage of energy legislation. However, until the climax of the Katrina disaster, the urgency of previous energy crises had been lacking. Throughout the period, U.S. gasoline consumption continued to rise. In part, this may be because although the price of gasoline in nominal terms set a record, in real terms it did not appear to be reaching the level of the Iranian crisis years of the early 1980s (see **Figure 6**); that is, until Katrina pushed it toward the $3.00-per-gallon mark. Further, unlike the earlier crises, there was no physical shortage of gasoline and there were no lines at the pump, except in local disaster-affected areas.

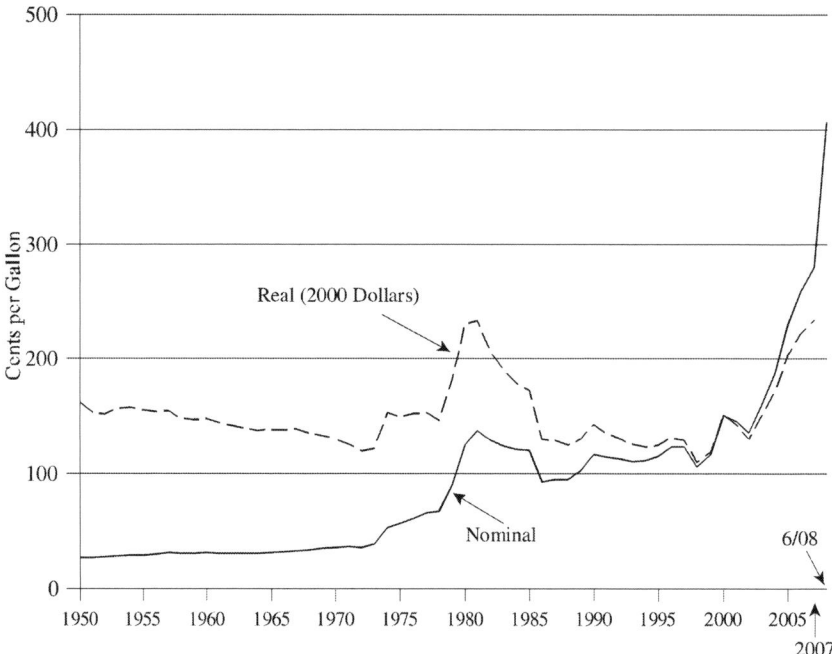

Source: EIA, *Annual Energy Review 2007*, Table 5.24 and *Monthly Energy Review*, July 2008, Table 9.4.

Figure 6. Nominal and Real Price of Gasoline, 1950-2007 and June 2008.

Gasoline Prices: Causes of Increases and Congressional Response   73

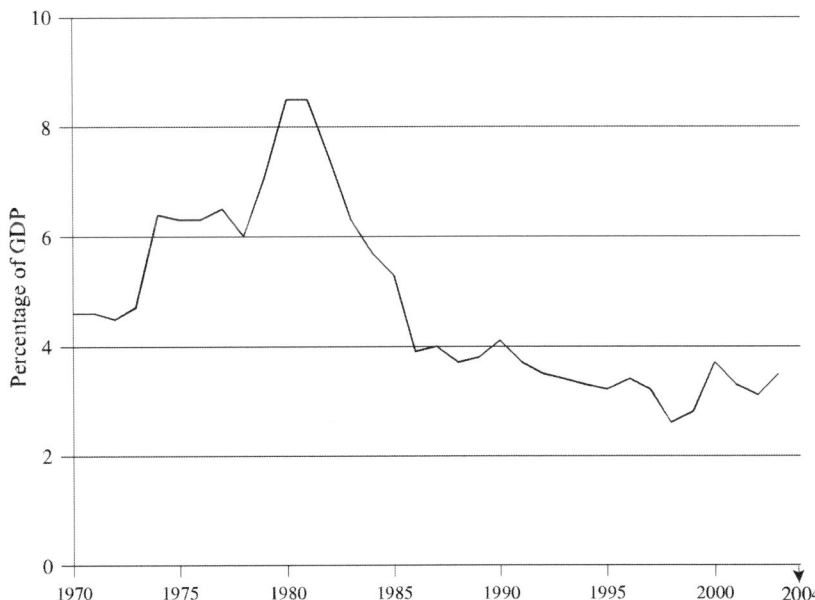

Source: EIA, Annual Energy Review, 2006, Tables 3.5 and D1

Figure 7. Consumer Spending on Oil as a Percentage of GDP, 1970-2004.

As **Figure 7** indicates, by the early 1990s the proportion of consumer expenditures on oil and gasoline had declined from the high levels of the 1970s and early 1980s. Data are not yet available to indicate what effect the price run-up starting in 2004 has had on this measure.

Perhaps most important, the common view during the earlier crises was that oil prices not only were high, but were destined to become ever higher in the coming years, because world resources were probably beginning to level off and would decline in the future. This view is no longer widely prevalent, largely because world proved reserves have increased faster than production, and are currently more than twice the level at the time of the Arab oil embargo in 1973.

At the beginning of the current crisis, the general expectation was that the price increase was a temporary phenomenon. In part, this may be due to the fact that there has been no physical shortage of gasoline or lines at the pump, as there were after the Arab oil embargo in 1973 and the Iranian revolution in 1979. But the continued and unrelenting increase in crude oil prices to record levels, even discounting inflation, is leading many to suggest that changing world market conditions may have led to permanent, or at least chronic, shortages of petroleum production capacity. The persistent increases in world demand for oil, despite higher prices, and the inability or unwillingness in many parts of the world, particularly in the Middle East, to develop known existing resources, appear to presage a continuing tight market, in which production capacity is only slightly greater than demand. Under those conditions, temporary interruptions in production, caused for example by local political crises or weather, are much more likely than normal to force prices upward.

Others continue to expect that growth in demand will moderate, and production will increase to meet demand, as it did following the shortages of the 1970s. They argue that the market price of oil appears to be much higher than production costs, and is being sustained by the expectation of continued strong demand in the indefinite future. In addition, they point to large profits flowing to oil producers, and political pressure to invest those profits in increased production.

Policy Options

Congress has considered numerous energy policy initiatives and enacted many of them. With the continuing pressure of rising prices, however, energy policy has once again become the focus of attention, both in the Congress and on the campaign trail.

Policy options include efforts to ameliorate the effects of high prices in the short term, and to attack the longer term problem. The latter options come in three major forms: to reduce consumption by increased efficiency without having a negative effect on the economy; to substitute alternative fuels at a cost comparable to the oil they replace; and to encourage production of more oil, either in this country or abroad.

The choice of these options depends to a certain extent on how the future of the oil market is viewed. Those who consider it likely that the present tightness of the market is likely to continue, as described above, tend to support alternative fuels and increased efficiency, and to denigrate efforts to increase oil production as futile and ineffective compared to the growth in world demand, which they expect to continue indefinitely. Those who view the present tightness of the world market as an aberration that can be relieved with adequate investment in new production capacity view any move to increase supply, anywhere in the world, as a positive signal that the tightness and volatility of the world oil market can be eased and prices can more closely reflect the cost of production.

Oil-Related Legislation

Two major bills were introduced in the Senate in May 2008, one by Senate Majority Leader Reid and the other by Senator Domenici, ranking Republican on the Energy and Natural Resources Committee. In addition, bills to suspend the federal gasoline tax during the 2008 summer driving season were introduced by Senator McCain and Senator Clinton. The gasoline tax suspension provision is not included in either Senator Reid's bill or Senator Domenici's bill.

Most but not all provisions of these and other bills described in this report are aimed at achieving one or more of the policy options described above. This section reviews major legislation that could affect the choice and cost of fuels for transportation, or affect the ability or motivation of industry to develop petroleum resources or alternative fuels, or the modes of transportation that use them. Issues that have a history of debate and legislation are also discussed in more detail.

Major Legislation. The main features of the gas tax moratorium and the Democratic and Republican Senate energy bills are described below.

Gas Tax Moratorium. Bills introduced by Senator McCain (S. 2890) and Senator Clinton (S. 2971) would suspend federal gasoline and diesel transportation taxes for the summer driving season, and the proposals have been a topic in the presidential campaigns of the two candidates. Senator Obama, also campaigning for the Democratic presidential nomination, has criticized Senator Clinton's proposal. Similar bills have been introduced in the House. (For details see CRS Report RL34475, *Transportation Fuel Taxes: Impacts of a Repeal or Moratorium,* by Robert Pirog and John W. Fischer.)

Consumer-First Energy Act of 2008 (S. 2991). S. 2991, introduced by Senator Reid and cosponsored by 22 Democratic Senators, has the following major provisions:

- A number of tax provisions affecting the oil and gas industry, related to the treatment of foreign profits, and also including a "windfall profits" tax on income in excess of "the reasonably inflated average profit" on crude oil;
- Creation of an "Energy Independence and Security Trust Fund," to be financed by funds received from the windfall tax provisions;
- A "Petroleum Consumer Price Gouging Protection" provision, similar to the price gouging protection proposals previously considered (see section on price gouging, below);

- Suspension of acquiring additional petroleum for the Strategic Petroleum Reserve (SPR) while the price of petroleum exceeds $75 per barrel (see section on SPR below);
- A "No Oil Producing and Exporting Cartels (NOPEC)" provision that would declare illegal collective action by foreign states to limit oil production, and deny sovereign immunity from prosecution to states that violated the provision;
- Set limits on speculation in energy commodities delivered in the United States in foreign boards of trade and require information regarding such speculative activity.

The American Energy Production Act of 2008 (S. 2958). S. 2958, sponsored by Senator Domenici with 21 Republican cosponsors, includes the following provisions:

- Allow oil and gas leasing in the Atlantic and Pacific Outer Continental Shelf (OCS), excluding the Gulf of Mexico, allowing governors of coastal states to petition for lifting the moratorium within their state boundaries, and creating a revenue-sharing plan in which states would receive 37.5% of revenue from new production (see section on OCS below);
- Establish oil and gas leasing in the coastal plain of the Arctic National Wildlife Refuge (ANWR: see below);
- Mandate production of 6 billion gallons of coal-derived fuel by 2022, to be produced without emission of greenhouse gas in excess of that emitted by the gasoline it replaces;
- Suspend filling the SPR for 180 days;
- Encourage commercial leasing of oil shale resources in Colorado, Wyoming and Utah.

On May 13, 2008, the Senate considered the provisions of the American Energy Production Act offered as an amendment to the Flood Insurance Reform and Modernization Act (S. 2284/H.R. 3121). The amendment was not adopted.

Reducing Impacts on Consumers. A number of proposals are aimed at easing the impact of high prices on consumers, or are aimed at the oil industry's price-making policies.

Price Gouging. The rapid increase in gasoline prices following the Katrina disaster led to allegations of price gouging. P.L. 109-58 included a provision requiring the Federal Trade Commission (FTC) to conduct an investigation into price gouging in increased gasoline prices.

The issue reemerged in the 110th Congress as gasoline prices surged past $3.00 per gallon. On May 23, 2007, the House passed the Federal Price Gouging Prevention Act (H.R. 1252). The bill would ban the sale of gasoline at "unconscionably excessive" prices during energy emergencies declared by the president, and impose heavy fines and imprisonment for violations. The White House complained that the bill could result in gasoline price controls, and threatened to veto it, but the House vote of 284-141 indicated enough support to override a veto.

The Senate, in passing its version of H.R. 6, the Creating Long-Term Energy Alternatives for the Nation (CLEAN Energy) Act of 2007 on June 21, 2007, included a price-gouging provision similar to that in H.R. 1252. However, the provision was not included in the final version of H.R. 6, which became P.L. 110-140.

The Consumer-First Energy Act (S. 2991) contains a provision on price gouging similar to the previously considered measures.

Speculation in the Oil Market. The possibility that speculation has unreasonably driven up oil prices, either because of illegal manipulation or because a speculative bubble is underway, has led to the introduction of much legislation to increase oversight or regulate speculation. The 2008 farm bill (Food Conservation and Energy Act, P.L. 110-234) contains provisions expanding the role of the Commodity Futures Trading Commission (CFTC), but many other bills are under consideration that would address other perceived problems. On June 26, 2008, the House passed the Energy Markets Emergency Act (H.R. 6377), which would direct the CFTC to curb excessive speculation, price distortion and other activity that is causing major market disturbances.

In the Senate, Majority Leader Reid proposed consideration of S. 3268, the Stop Excessive Energy Speculation Act of 2008, but disagreement on how to treat amendments on the floor blocked action. A similar bill, the Commodity Markets Transparency and Accountability Act of 2008 (H.R. 6604), was approved by the House Agriculture Committee July 24, 2008, but failed to get 2/3 of the vote under suspension of the rules on the House floor July 30. (For details, see CRS Report RL34555, *Speculation and Energy Prices: Legislative Responses*, by Mark Jickling and Lynn J. Cunningham.)

Filling the Strategic Petroleum Reserve. Authorized in 1975, SPR consists of caverns formed out of naturally occurring salt domes in Louisiana and Texas in which nearly 700 million barrels of crude oil are stored. Its current capacity is 727 million barrels, and it is authorized at 1 billion barrels. The purpose of the SPR is to provide an emergency source of crude oil that may be tapped in the event of a presidential finding that an interruption in oil supply, or an interruption threatening adverse economic effects, warrants a drawdown from the reserve.

Program costs for the SPR in recent years have been dedicated principally to maintaining SPR facilities and keeping the SPR in readiness should it be needed. Since FY1999, any fill of the SPR has been with deliveries of royalty-in-kind (RIK) oil to the SPR in lieu of cash royalties to the federal government on offshore production. Through FY2007, royalty-in-kind deliveries to the SPR have totaled roughly 140 million barrels and forgone receipts to the Department of the Interior are estimated to be $4.6 billion. DOE has projected deliveries of RIK oil during FY2008 of 19.1 million barrels and $1.170 billion in forgone revenues.

Continued fill of the SPR with royalty-in-kind oil has been controversial. Critics argue that it is inadvisable to add oil to the SPR when markets are tight and prices remain high. They argue further that the additional oil adds little to U.S. energy security. Supporters of RIK fill argue that the fill rate is too little to have a discernible impact on markets, and that currently high refined-product prices are sustained by factors other than crude supply, which is more than ample at this time.

Legislation was introduced in the Second Session (H.R. 5146, S. 2598) to suspend RIK fill. The House bill would also mandate a sale of 13 million barrels of SPR oil during FY2008, with the proceeds to be spent on a number of energy efficiency and alternative fuel programs. Both bills would establish conditions, including a significant decline in crude oil prices, that would have to be satisfied before RIK fill could be resumed. On May 13, 2008, the House passed, 385-25, a similar bill, the Strategic Petroleum Reserve Fill Suspension and Consumer Protection Act (H.R. 6022), which would suspend SPR fill until the end of 2008 unless the price of oil dropped below $75 per barrel. The Senate passed the bill on the same day, and it became P.L. 110-232 on May 19.

A further move regarding the SPR was the Consumer Energy Supply Act of 2008 (H.R. 6578), introduced by Representative Lampson. The bill would have required the President to release 70 million barrels of high-quality crude oil from the SPR over the next six months, a move which proponents said would lower gasoline prices. The House took up the bill on July 24, 2008, under suspension of the rules, which required a 2/3 vote in favor for passage. The vote was 268-157, not enough to pass the bill. (For details see CRS Report RL33341, *The Strategic Petroleum Reserve: History, Perspectives, and Issues*, by Robert Bamberger).

Mid- to Long-Term Supply and Demand. Most proposals affecting supply and demand of crude oil and gasoline would not affect the current short-term crisis but would be aimed at longer term trends.

Fuel Economy. Corporate average fuel economy (CAFE) standards also have a long history of controversy, going back to their establishment in the 1970s. In the mid-1990s, the National Highway Traffic Safety Administration (NHTSA) was considering a rulemaking that would result in increased standards for light-duty trucks (including sport utility vehicles), but for several years, Congress included in its annual appropriation for NHTSA a measure prohibiting NHTSA from analyzing or undertaking such a ruling. That prohibition was dropped in the FY2004 NHTSA appropriations, and a final rule issued by NHTSA in April 2003 required a boost in light-truck fuel economy to 22.2 miles per gallon by model year 2007. New fuel economy standards for light trucks were issued in 2006, to take effect in the 2008 model year, but implementation was blocked in court.

During House floor debate on P.L. 109-58, an amendment to increase fuel economy standards to 33 miles per gallon over 10 years was defeated by a vote of 177-254. A more general amendment to the House bill, requiring the Administration to take "voluntary, regulatory, and other actions" to reduce oil demand in the United States by 1 million barrels per day from projected levels by 2013 was defeated 166- 262. The measure was included in the bill passed by the Senate but was dropped in conference.

Continued high gasoline prices raised congressional interest in higher mandated CAFE standards again. On January 22, 2007, a bipartisan group of 10 Senators introduced S. 357, the Ten-in-Ten Fuel Economy Act, which would raise standards for SUVs and passenger cars to 35 mpg by 2019. The President argued that standards should be set by the executive branch, not by Congress, and in his State of the Union speech on January 23, 2007, he set a goal of reducing gasoline consumption by 5% by 2017 through more stringent standards. The White House said that would be the equivalent of increasing CAFE standards 4% per year starting with model year 2010.

After considerable debate, P.L. 110-140 was passed and signed in December 2007, including setting a target of 35 miles per gallon for the combined fleet of cars and light trucks by model year 2020. A number of new procedures, including the trading of fuel economy credits among auto manufacturers, were included in the bill. (For details see CRS Report RL33413, *Automobile and Light Truck Fuel Economy: The CAFE Standards,* by Brent D. Yacobucci and Robert Bamberger.)

ANWR. Oil and gas exploration and development of part of the Arctic National Wildlife Refuge (ANWR) has been controversial for many years. This was part of the early proposals for legislation that eventually became the Energy

Policy Act of 1992 (P.L. 102-486), but was dropped in the face of strong opposition in both houses. Support for the action grew gradually in the following years, along with technological developments that advocates claimed would reduce the environmental impact of development. Numerous attempts to open the region for leasing have been made, and both the House and the Senate at various times approved measures that included leasing provisions, but none of them have survived to become law. (For more details, see CRS Report RL32838, *Arctic National Wildlife Refuge (ANWR): Legislative Actions Through the 110th Congress, First Session*, by Anne Gillis, M. Lynne Corn, and Elizabeth A. Roberts.)

Savings Goals. A number of legislative proposals would have set goals for reducing oil consumption. An example is the Enhanced Energy Security Act of 2006 (S. 2747), introduced by Senator Bingaman May 4, 2006, which would have required the Director of the Office of Management and Budget to develop an action plan to save 2.5 million barrels per day (mbd) in 2016, 7 mbd in 2026, and 10 mbd in 2031.

President Bush took up the idea in his State of the Union speech on January 23, 2007, calling for a cut in gasoline consumption of 20% in 10 years, through a combination of increased fuel economy standards (see above) and increased mandated use of alternative fuels (see below).

Alternative Fuels. In his January 31, 2006 State of the Union message, President Bush asserted that the United States is "addicted to oil," and set the goal of replacing more than 75% of oil imports from the Middle East by 2025. The main thrust of the presidential initiative was to increase funding for research in producing ethanol from plant fiber biomass (rather than from corn), for improved batteries for hybrid automobiles, and for hydrogen fuels.

In his next State of the Union speech, on January 23, 2007, the President went further, setting a goal of reducing gasoline consumption by 20% in 10 years, through a combination of more stringent fuel economy standards and setting a mandatory renewable fuels standards of 35 billion gallons of renewable and alternative fuels by 2017, about five times the current consumption. The Energy Policy Act of 2005 (P.L. 109-58) set a target of 7.5 billion gallons by 2012.

On June 21, 2007, the Senate passed its version of H.R. 6, the Creating Long-Term Energy Alternatives for the Nation (CLEAN Energy) Act of 2007, including a provision requiring production of 36 billion gallons of ethanol in 2022. The final version of the bill, P.L. 110-140, set a modified standard that starts at 9.0 billion gallons of renewable fuel in 2008 and rises to 36 billion gallons by 2022. Of the latter total, 21 billion gallons is required to be obtained from cellulosic ethanol and other advanced biofuels. (For more details, see CRS Report RL34265,

Selected Issues Related to an Expansion of the Renewable Fuel Standard (RFS), by Brent D. Yacobucci and Randy Schnepf.)

OCS Leasing. The moratorium on oil and gas leasing in the Outer Continental Shelf (OCS), except in the central and western Gulf of Mexico and some parts of Alaska, was subject to much controversy during consideration of P.L. 109- 58. A proposal to allow states to voluntarily opt out of the moratorium was dropped under threat of filibuster, and a measure to order the Department of the Interior to perform an inventory of OCS resources barely survived the debate.

Following the disruption of production by Hurricane Katrina, momentum to lift the moratorium increased, along with efforts by Gulf states to increase their share of revenues from oil and gas production in the Gulf of Mexico. This movement culminated in S. 3711, the Gulf of Mexico Energy Security Act of 2006, which lifted some restrictions in Gulf of Mexico oil and gas leases and increased revenue sharing for Gulf producing states. The Senate passed S. 3711 on August 1, 2006, and its provisions were included in H.R. 6111 (P.L. 109-432), the Tax Relief and Health Care Act of 2006, which passed the House on December 8 and the Senate the following day. (For details, see CRS Report RL3 3493, *Outer Continental Shelf: Debate Over Oil and Gas Leasing and Revenue Sharing*, by Marc Humphries.)

Representative Barton's proposed energy bill, which he submitted in the form of a motion to recommit H.R. 3221 on August 4, 2007, included a provision to open up the OCS to oil and gas leasing beyond 100 miles of the coast. The motion to recommit was defeated by a vote of 169 ayes to 244 noes.

Another issue regarding OCS leasing concerns a number of leases issued in 1998 and 1999 which granted royalty relief under certain conditions without including a price threshold. Several initiatives to force renegotiation of these contracts have been proposed, including the House-passed version of H.R. 6, the CLEAN Energy Act. Similar provisions, including denial of new Gulf of Mexico leases to lessees holding leases without price thresholds, and establishing "conservation of resources" fees, were included in H.R. 3221, as passed by the House August 4, 2007. However, the provision was not included in the final bill, P.L. 110- 140. (For details on OCS royalty relief issues see CRS Report RS22567, *Royalty Relief for U.S. Deepwater Oil and Gas Leases*, by Marc Humphries.)

LEGISLATION

H.R. 1596 (Ferguson). Clean and Green Renewable Energy Tax Credit Act of 2007.

H.R. 2419 (Peterson). Food Conservation and Energy Act of 2008. Contains provisions expanding the role of the Commodity Futures Trading Commission for certain energy derivatives. Enacted May 22, 2008, over the President's veto (P.L. 110-234).

H.R. 2448 (Kuhl). Emergency Gas Price Relief Act of 2007.

H.R. 5146 (Lampson). Invest in Energy Security Act. Would suspend SPR fill, and sell SPR oil to finance an Energy Independence and Security Fund.

H.R. 6022 (Welch). Strategic Petroleum Reserve Fill Suspension and Consumer Protection Act. Passed the House and the Senate May 13, 2008. Became P.L. 110-232 on May 19, 2008.

H.R. 6349 (Marshall). Increasing Transparency and Accountability in Oil Prices Act of 2008.

H.R. 6377 (Peterson). Energy Markets Emergency Act of 2008. Passed the House June 26, 2008.

H.R. 6578 (Lampson). Consumer Energy Supply Act of 2008. Would release oil from the Strategic Petroleum Reserve over the next six months. On July 24, 2008, in the House, on motion to suspend the rules and pass the bill, as amended, failed by the Yeas and Nays (2/3 required): 268-157.

H.R. 6604 (Peterson). Commodity Markets Transparency and Accountability Act of 2008. On July 30, 2008, in the House, on motion to suspend the rules and pass the bill, as amended, failed by the Yeas and Nays: (2/3 required): 276-151.

S. 2598 (Dorgan). Strategic Petroleum Reserve Fill Suspension and Consumer Protection Act of 2008.

S. 2896 (Snowe). Diesel Tax Parity Act of 2008.

S. 2958 (Domenici). American Energy Production Act of 2008.

S. 2971 (Clinton). A bill to amend the Internal Revenue Code of 1986 to provide for a suspension of the highway fuel tax, and for other purposes.

S. 2991 (Reid). Consumer-First Energy Act of 2008.

S. 3268 (Reid). Stop Excessive Energy Speculation Act of 2008. Cloture motion on the motion to proceed to the measure presented in Senate 7/17/2008.

ENDNOTES

[1] International Energy Agency, *Oil Market Report,* April 11, 2008, p. 1.

[2] Energy Information Administration data based on March 2008 data and a base price of gasoline of $3.24 per gallon. See Gasoline & Diesel Fuel Update [http://www.eia.doe.gov].

[3] The price of diesel fuel for transportation has also surged to record levels. For details on the relationship between diesel and gasoline prices, see CRS Report RL34431, The Disparity Between Retail Gasoline and Diesel Fuel Prices, by Robert Bamberger and Robert Pirog.

[4] Energy Information Administration. Weekly Petroleum Status Report, August 29, 2008, p. 25.

CHAPTER SOURCES

The following chapters have been previously published:

Chapter 1 is an edited, reformatted and augmented version of a Congressional Research Service publication, Report RL34625, dated August 20, 2008.

Chapter 2 is an edited, reformatted and augmented version of a Congressional Research Service publication, Report RL34431, dated March 31, 2008.

Chapter 3 is an edited, reformatted and augmented version of a Congressional Research Service publication, Report RL34475, dated May 7, 2008.

Chapter 4 is an edited, reformatted and augmented version of a United States Government Accountability Office publication GAO-08-1114, dated September 2008.

Chapter 5 is an edited, reformatted and augmented version of a Congressional Research Service publication, Report RL33521, updated September 25, 2008.

INDEX

A

accuracy, 42, 47, 54, 56
additives, 54
adjustment, 3
administrative, 50
advocacy, 45
agriculture, vii, viii, 1, 2, 10
aid, 37, 42, 46, 51, 52, 53, 54, 58, 79
air, 71
air pollution, 71
Alaska, viii, 2, 14, 81
alcohol, 37, 68
Algeria, 13
alternative, viii, 2, 29, 38, 48, 74, 75, 78, 80
alternative energy, viii, 2
amendments, 77
analysts, 11, 35
Angola, 13
appendix, 45, 60
appropriations, 67, 79
Arabia, 11, 12, 13
Arctic, 67, 68, 76, 79
Arctic National Wildlife Refuge, 67, 68, 76, 79
Arizona, 62
Asphalt, 20
assumptions, 38, 46, 55
Atlantic, viii, 2, 14, 76
auditing, 45, 62
Australia, 45, 56, 58, 62
authority, 38
automobiles, viii, 2, 66, 68, 80
availability, 27, 35, 54
averaging, 19
aviation, 16

B

back, 12, 79
barges, 48
batteries, 80
behavior, 8, 9, 15, 24
Belgium, 43, 45, 47, 56, 57, 59, 62
benefits, vii, xii, 42, 46, 47, 51, 53, 54, 56, 59, 60, 65
benign, 39
bias, 11
biodiesel, 48
biofuels, 68, 80
biomass, 80
bipartisan, 79
blends, 48, 54, 55, 71
boils, 62
bonds, 10
broad spectrum, 67
bubble, 10, 77
budget deficit, x, 31
Bush Administration, 38
buyer, 9

C

calibration, 56
campaigns, x, xi, 31, 32
Canada, 45, 47, 56, 57, 59, 60, 62, 63
candidates, 68, 75
carbon, 15, 36
carbon emissions, 15, 36
cash flow, 4
cellulosic, 68, 80
cellulosic ethanol, 80
CFTC, 7, 77
chemicals, 62
Chevron, 35
China, 5, 6, 11, 25, 69
classes, 53
CNN, 16
coal, 76
codes, 50
Coke, 20
Colorado, 76
Committee on Homeland Security, 10, 16
commodities, 10, 76
commodity, vii, viii, xii, 2, 7, 9, 10, 28, 65, 69
commodity futures, vii, xii, 10, 65
Commodity Futures Trading Commission (CFTC), 77, 82
commodity markets, 10, 11, 69
community, 37, 39, 54, 59
compensation, xi, 41, 42, 44, 45, 46, 47, 48, 50, 51, 52, 53, 54, 55, 56, 57, 58, 59, 60, 61, 63
competition, 20, 34
components, 19, 22, 26, 48, 62
composition, 3, 28, 48
conception, 5, 9
confidence, 28
Congress, vii, viii, x, xi, xii, 2, 14, 31, 32, 33, 34, 39, 44, 65, 66, 67, 68, 74, 77, 79, 80
Congressional Budget Office, 37
congressional hearings, 9
conjecture, 25
consensus, viii, 2, 44, 51
conservation, viii, 2, 81
constraints, 12
construction, 14
consumer expenditure, 73
consumer goods, x, 18
consumer protection, 50
consumers, vii, viii, x, xi, 1, 2, 3, 5, 9, 10, 11, 12, 14, 18, 24, 27, 28, 31, 33, 34, 35, 36, 42, 45, 46, 47, 48, 50, 51, 52, 53, 54, 55, 56, 57, 58, 59, 60, 68, 71, 76
consumption, 3, 5, 6, 12, 15, 16, 19, 36, 37, 48, 66, 69, 72, 74, 79, 80
contamination, 27
content analysis, 61
continental shelf, viii, 2, 14
contracts, 9, 10, 38, 43, 81
control, 5, 7, 9, 15, 47
conversion, 58, 63
corn, 80
Corporate Average Fuel Economy, 66
correlation, 9
cost-benefit analysis, 45, 47, 56, 59, 62
costs, viii, ix, x, xi, 2, 12, 15, 17, 18, 19, 22, 24, 27, 28, 29, 32, 33, 37, 39, 41, 42, 44, 45, 46, 47, 51, 52, 53, 54, 55, 56, 59, 60, 71, 74, 78
costs of production, 28
CRS, 16, 23, 24, 39, 67, 70, 75, 77, 78, 79, 80, 81, 83
crude oil, vii, viii, ix, xii, 1, 2, 3, 4, 5, 6, 7, 8, 9, 12, 17, 19, 20, 24, 25, 26, 28, 32, 33, 35, 36, 48, 55, 62, 65, 66, 68, 69, 71, 73, 75, 77, 78, 79
currency, 11

D

debates, viii, 2
decisions, ix, x, 9, 10, 11, 18, 37, 46, 47, 53, 55, 59, 60
deficit, xi, 32, 39
delivery, 8, 9, 28
Democrats, 68
denial, 81
density, 48, 56
Department of Commerce, 44

Department of Energy, 20, 21, 22, 23, 24, 29, 39, 66
Department of the Interior, 78, 81
deposits, 14, 34, 68
derivatives, 82
detection, 50
diesel, ix, x, xi, 17, 18, 20, 21, 22, 24, 25, 26, 27, 28, 29, 32, 33, 34, 35, 36, 41, 43, 44, 45, 48, 51, 57, 61, 62, 68, 75, 83
diesel fuel, ix, x, 17, 18, 20, 22, 24, 25, 26, 27, 28, 29, 32, 33, 35, 36, 83
disaster, 72, 76
disclosure, 46, 52
discounting, xii, 66, 73
disposable income, 33
disposition, 50
distillates, ix, 17, 19, 20, 21, 28
distillation, 19
distribution, viii, ix, 1, 18, 22, 27, 39, 48, 60
diversification, 10
domestic demand, 12, 26
domestic markets, 12
Dow Jones Industrial Average, 10

E

economic growth, vii, viii, 1, 2, 12, 15, 25, 72
economic performance, 4
economic theory, 9, 42, 46, 55, 60
Education, 51
Egypt, 11
elasticity, 39
elasticity of supply, 39
electric power, 21
embargo, xii, 66, 73
emission, 15, 76
employment, 14
end-users, 35
energy, vii, viii, xi, xii, 2, 7, 10, 14, 41, 42, 43, 46, 47, 48, 53, 54, 57, 65, 68, 69, 72, 74, 75, 76, 77, 78, 81, 82
energy efficiency, 78
Energy Independence and Security Act, vii, xi, 65, 66, 68

Energy Information Administration (EIA), 3, 4, 8, 16, 20, 21, 22, 23, 24, 29, 39, 82, 83
energy markets, 10
Energy Policy Act, vii, xi, 65, 67, 80
Energy Policy Act of 2005, vii, xi, 65, 67, 80
environment, 5, 35
environmental impact, 80
Environmental Protection Agency (EPA), 27, 45, 50, 61
equity, 42, 47, 54, 56
estimating, 39
ethanol, 48, 54, 71, 80
Europe, ix, 16, 18, 25
European Union (EU), 56, 58
evolution, x, 18
excess supply, 69
exchange rate, 11
excise tax, viii, 2, 4, 14, 33, 34, 35, 36, 39
exercise, 5
expertise, 13, 60
exports, 12
extraction, 12

F

fairness, 57
February, 3, 4, 8, 39, 51
federal budget, x, 31, 37
federal government, 47, 58, 66, 78
Federal Highway Administration, 16, 38
Federal Trade Commission (FTC), 45, 50, 61, 77
feedstock, ix, 17, 19, 22
fees, 53, 81
fiber, 80
finance, 39, 82
financial difficulty, 39
financial institution, 9
financial institutions, 9
financial markets, 8
financing, 39
fines, 77
firms, 5, 35
fixed costs, 12
flow, 4, 26

fuel, viii, ix, x, xi, 2, 11, 12, 16, 17, 18, 19, 20, 22, 24, 25, 26, 27, 28, 31, 32, 33, 34, 35, 36, 37, 38, 39, 41, 42, 43, 45, 46, 47, 48, 49, 50, 51, 52, 53, 54, 55, 56, 57, 58, 59, 60, 62, 63, 71, 76, 78, 79, 80, 82
fuel type, 33
fuel-efficient vehicles, 37
funding, xi, 13, 31, 33, 36, 37, 38, 80
funds, 7, 9, 10, 33, 36, 39, 42, 46, 53, 58, 75
futures, vii, viii, xii, 2, 7, 8, 9, 10, 14, 15, 32, 35, 65
futures markets, vii, viii, xii, 2, 7, 8, 9, 14, 35, 65

G

gas, vii, xi, xii, 4, 29, 38, 41, 42, 43, 45, 46, 48, 54, 65, 67, 68, 75, 76, 79, 81
gas exploration, 79
gasoline, vii, viii, ix, x, xi, xii, 1, 2, 3, 4, 5, 9, 11, 12, 14, 15, 16, 17, 18, 19, 22, 24, 25, 26, 27, 28, 29, 32, 33, 34, 35, 36, 39, 41, 43, 44, 45, 48, 49, 51, 55, 57, 61, 62, 65, 66, 67, 68, 71, 72, 73, 75, 76, 77, 78, 79, 80, 82, 83
GDP, 73
general fund, 33, 38, 39
generation, 21
Georgia, 15
Global Insight, 39
global warming, 15
Globalization, 62
goals, viii, 2, 15, 80
government, iv, xi, 12, 15, 41, 43, 44, 45, 47, 50, 56, 57, 58, 59, 60, 62, 66, 78
government policy, 15
governors, 76
grades, x, 18, 32
Great Britain, 47
greenhouse, 48, 76
greenhouse gas, 48, 76
groups, xi, 41, 42, 45, 46, 50, 53, 61, 62
growth, vii, viii, xii, 1, 2, 6, 7, 10, 12, 15, 20, 24, 25, 35, 37, 66, 72, 74
growth rate, 6

guidance, 44, 57
Gulf of Mexico, vii, viii, xii, 2, 14, 65, 76, 81

H

hands, 7
harm, 57
Hawaii, 42, 47, 56, 59, 60, 62
health, 48
health problems, 48
heat, 19
heating, ix, x, 17, 18, 19, 20, 27, 29, 32, 48, 57, 62
heating oil, 20, 27, 32, 62
hedge funds, 9
hedging, 7, 16, 35
highways, 36, 37
home heating oil, ix, x, 17, 18, 19, 20, 27, 48, 57
Homeland Security, 10, 16
horizon, 10
House, 34, 43, 66, 67, 68, 75, 77, 78, 79, 80, 81, 82
Hurricane Katrina, 81
hybrid, 80
hybrids, 37
hydro, 19
hydrocarbon, xi, 41, 43
hydrocarbon fuels, xi, 41, 43
hydrocarbons, 19
hydrogen, 80

I

ICE, 7
id, vii, 1, 3, 36
immunity, 76
impact energy, 54
implementation, 27, 45, 46, 50, 52, 53, 54, 56, 58, 61, 79
importer, 39
imports, 12, 15, 19, 26, 28, 80
imprisonment, 77
incentive, 3, 28, 35

incentives, 6, 36
income, 12, 33, 35, 36, 75
India, 5, 6
Indonesia, 11, 13
industrial, x, 18, 25
industrial application, x, 18
industry, vii, xi, 1, 6, 7, 15, 26, 34, 35, 42, 43, 45, 46, 47, 48, 50, 53, 55, 56, 59, 61, 75
inelastic, 5, 28
infinite, 39
inflation, vii, viii, x, xii, 1, 2, 18, 35, 66, 73
infrastructure, 48
injury, iv
inspection, 42, 46, 52, 53, 57
inspections, 53
inspectors, 52, 59
interaction, 69
Internal Revenue Code, 82
International Energy Agency, 82
international markets, 27
interview, 45, 61
interviews, 45, 61
inventories, 19, 35, 47
inventory records, 54
investment, ix, x, 9, 10, 12, 15, 17, 18, 19, 27, 28, 46, 53, 54, 74
investors, 8, 9, 10
Iran, 11, 12, 13
Iraq, vii, xii, 13, 65
isolation, 6

J

Japan, 5, 16
jet fuel, 62
judgment, 42, 46, 55, 60
jurisdiction, 50
jurisdictions, 56
justification, 15

K

Katrina, 69, 72, 76, 81
Kuwait, 13

L

labor, 6
labor force, 6
land, 53
law, 20, 50, 80
laws, 43, 50, 59
leaks, 46, 54
legislation, viii, 2, 14, 38, 44, 52, 58, 59, 61, 66, 67, 68, 72, 75, 77, 79
legislative proposals, 68, 80
Libya, 13
lift, 81
light trucks, 37, 66, 68, 79
likelihood, ix, 2, 26
linkage, ix, 2, 9, 10
liquids, 48
livestock, 10
local government, 50
location, 57
loopholes, 14
losses, 7, 57
Louisiana, 77
lower prices, 12, 16, 34
LUST, 33, 34, 36

M

magnetic, iv
Maine, 62
maintenance, 19, 51
management, 42, 60
manipulation, 14, 77
manufacturer, 51
manufacturing, 6
market, viii, ix, x, xii, 2, 3, 5, 6, 7, 8, 9, 10, 11, 12, 13, 14, 15, 18, 19, 20, 21, 25, 26, 27, 28, 31, 34, 35, 46, 55, 58, 66, 69, 70, 71, 73, 74, 77
market concentration, 55
market economy, 28
Market forces, 15
market prices, 10, 69
marketing, viii, xi, 1, 22, 39, 41, 68, 71

marketplace, 45, 50, 51, 54, 61
markets, vii, viii, x, xii, 2, 7, 8, 9, 10, 12, 14, 18, 19, 26, 27, 32, 35, 65, 78
Massachusetts, 62
measurement, 42, 43, 47, 51, 54, 56, 57, 58
measures, viii, ix, xi, 2, 15, 34, 36, 41, 44, 45, 46, 50, 54, 56, 57, 58, 59, 62, 68, 77, 80
mergers, 55
metals, 10
Mexico, 81
Middle East, 69, 74, 80
military, 21
million barrels per day, 5, 12, 26, 35, 66, 79, 80
Minnesota, 49, 62
misconception, 49
misleading, 6
missions, 48
Missouri, 52, 59, 62
mixing, 20
momentum, 81
Montana, 62
moratorium, viii, x, xi, 2, 31, 32, 33, 38, 67, 75, 76, 81
motion, 81, 82
motivation, 75
movement, 11, 81
multiplicity, viii, 2

nation, 50, 52
National Highway Traffic Safety Administration (NHTSA), 79
National Institute of Standards and Technology (NIST), xi, 41, 44, 60, 61
National Wildlife Refuge, viii, 2, 67, 68, 76, 79
nationalism, 12
natural, 11, 69
natural disasters, 69
Nebraska, 62
Netherlands, 62
New York, iii, iv, 7, 39, 62
New York Mercantile Exchange, 7

New York Times, 39
NHTSA, 79
Nigeria, 13, 15
NIST, xi, 41, 44, 45, 49, 50, 61, 62
normal, 34, 52, 57, 74
North America, 52
nuclear, 12
NYMEX, 7, 9, 15

obligation, 38
octane, 62
Office of Management and Budget, 37, 39, 80
Offices of Congressional Relations and Public Affairs, 60
offshore, viii, 2, 78
oil, vii, viii, ix, x, xi, xii, 1, 2, 3, 4, 5, 6, 7, 8, 9, 10, 11, 12, 13, 14, 15, 17, 18, 19, 20, 24, 25, 26, 27, 28, 29, 32, 33, 34, 35, 36, 42, 45, 46, 48, 53, 55, 57, 62, 65, 66, 67, 68, 69, 71, 73, 74, 75, 76, 77, 78, 79, 80, 81, 82
oil production, 6, 12, 13, 35, 74, 76
oil refining, 15
oil shale, 68, 76
Oklahoma, 7
OMB, 37, 38
operator, 39
opportunity costs, 29
opposition, xi, 32, 80
Oregon, 62, 63
Organization of the Petroleum Exporting Countries (OPEC), 5, 12, 13, 14, 70
OTC, 7
oversight, 14, 16, 77
ozone, 48

P

Pacific, viii, 2, 14, 76
partnership, 44
passenger, 79
Pennsylvania, 62
pension, 9

perception, 70
Persian Gulf, 5
petroleum, viii, ix, xi, xii, 2, 5, 12, 13, 14, 15, 16, 17, 19, 20, 25, 26, 27, 29, 32, 33, 34, 35, 41, 44, 45, 47, 48, 49, 50, 55, 57, 60, 61, 62, 66, 68, 71, 73, 75, 76, 77, 78, 82, 83
petroleum products, ix, 15, 17, 19, 26, 32, 33, 34, 35, 47, 48, 49, 50, 57, 62
Philadelphia, 16
pipelines, 48
planning, 16
play, 24
policy initiative, 74
policymakers, 42, 47, 55
pollution, 71
population, 11
portfolio, 7, 10
portfolios, 10
power, 21
power generation, 21
powers, 14
prediction, 25, 37
premium, 21
president, 77
President Bush, 80
presidential campaigns, 75
pressure, vii, viii, xii, 1, 4, 15, 19, 25, 26, 35, 36, 37, 65, 68, 69, 71, 74
price changes, 12, 34
price effect, ix, 7, 14, 18, 25
price gouging, vii, xii, 65, 75, 76, 77
price manipulation, 14
price-gouging, 77
prices, vii, viii, ix, x, xi, xii, 1, 2, 3, 5, 6, 7, 8, 9, 10, 11, 12, 13, 14, 15, 16, 17, 18, 19, 20, 21, 22, 24, 25, 26, 27, 28, 29, 31, 32, 33, 34, 35, 36, 42, 43, 46, 53, 55, 58, 60, 65, 66, 67, 68, 69, 70, 71, 73, 74, 76, 77, 78, 79, 83
private, 9, 50
private sector, 50
producers, vii, xii, 3, 5, 12, 25, 65, 74
product market, 34

production, vii, ix, xii, 1, 5, 6, 7, 12, 13, 15, 17, 19, 24, 25, 26, 28, 29, 66, 68, 70, 73, 74, 76, 78, 80, 81
production costs, 74
productive capacity, 12
profit, 26, 35, 55, 75
profit margin, 35
profitability, 24
profits, vii, ix, 1, 2, 26, 33, 34, 35, 68, 74, 75
program, 37
property, iv
protection, 50, 75
public, viii, 2, 26, 50
pumps, 42, 43, 45, 48, 50, 51, 52, 55, 56, 57, 59, 61, 62, 63

Q

Qatar, 13

R

range, ix, 5, 17, 19, 25, 26, 45, 51, 55, 61, 71
raw material, 29
raw materials, 29
real terms, 72
refineries, 25, 55
refiners, ix, x, 4, 14, 17, 18, 19, 24, 27, 32, 34, 36
refinery capacity, vii, xii, 65
refining, viii, ix, 1, 4, 12, 15, 17, 19, 21, 22, 24, 27, 28, 35, 39, 49, 55, 62, 71
reflection, 22
regional, x, 18, 50, 68
regular, 16, 62
regulation, 7, 16, 27, 50, 58
regulations, xi, 20, 41, 43, 44, 50, 58, 59, 62
reimbursement, 34, 38
relationship, x, 18, 34, 83
relative prices, ix, x, 17, 18, 24, 25, 26, 28
remodeling, 42, 46, 53
Renewable Fuel Standard (RFS), 81
Republican, 75, 76
reserves, 5, 7, 13, 73

residential, x, 18, 20, 21
resolution, 59
resources, 12, 13, 14, 73, 74, 75, 76, 81
respiratory, 48
responsiveness, 34, 39
retail, x, xi, 3, 19, 22, 32, 33, 34, 41, 42, 43, 44, 45, 46, 48, 49, 50, 51, 52, 53, 54, 55, 56, 57, 58, 59, 60, 61
returns, 21
revenue, 14, 36, 37, 38, 76, 81
RFS, 81
risk, 7, 10
Rita, 69
royalties, 78
royalty, 78, 81
rural, 53
rural areas, 53
Russia, 11, 15
rust, 33

S

SAFETEA-LU, 37
sales, 20, 22, 38, 42, 44, 45, 48, 51, 54, 56, 57, 58, 60
salt, 77
salt domes, 77
sampling, 61
Saudi Arabia, 11, 12, 13
search, xii, 44, 61, 66
seasonality, ix, 18, 21, 36
security, 78
Senate, 10, 16, 34, 66, 68, 75, 76, 77, 78, 79, 80, 81, 82
Senator Reid, 68, 75
sensitivity, 13, 34
series, 55
services, iv, x, 18, 28
shareholder value, 28
shares, 10
sharing, 76, 81
Shell, 35
shipping, x, 18, 25
shock, 58
short run, 12

shortage, xii, 66, 72, 73
shortages, xii, 5, 26, 52, 66, 73, 74
short-term, 79
signs, 46, 52
similarity, 20
singular, 5
sites, 29
South Dakota, 62
South Korea, 5
spare capacity, 6, 12, 16, 70
spectrum, 67
speculation, vii, viii, xii, 2, 14, 35, 65, 68, 76, 77
speech, 79, 80
speed, 63
spot market, 8, 9, 10
SPR, 66, 76, 77, 78, 82
stability, viii, 2
stakeholder, 45, 46, 53, 55, 61, 62
stakeholder groups, 61
stakeholders, xi, 41, 42, 44, 45, 46, 51, 52, 53, 54, 55, 58, 60, 61
standards, xi, 41, 44, 45, 46, 47, 50, 54, 57, 60, 61, 62, 66, 68, 79, 80
State of the Union, 79, 80
stock, 6
storage, 36, 48, 49, 50, 54, 62
Strategic Petroleum Reserve, viii, ix, 2, 14, 66, 76, 77, 78, 82
strategies, 7, 14
subsidies, 11, 37
subsidization, viii, 2
subsidy, 12
substances, 47
sulfur, ix, x, 18, 19, 20, 27
summer, viii, x, xi, 2, 19, 25, 31, 32, 33, 38, 39, 66, 68, 75
supply, vii, viii, xii, 1, 2, 5, 6, 8, 10, 12, 14, 16, 25, 26, 28, 29, 34, 39, 48, 49, 65, 66, 68, 69, 74, 78, 79
supply chain, viii, 1, 5
supply disruption, vii, xii, 7, 65, 69
surpluses, 5

Index

T

tanks, 48, 49, 50, 54, 62
tax collection, x, xi, 31, 32
tax cuts, 35
tax rates, 33
tax receipt, 53
Tax Relief and Health Care Act, 81
taxation, 36
taxes, x, xi, 5, 20, 22, 31, 32, 33, 34, 35, 36, 38, 39, 55, 67, 71, 75
technological developments, 80
temperature, xi, 19, 41, 42, 43, 44, 45, 46, 47, 48, 49, 50, 51, 52, 53, 54, 55, 56, 57, 58, 59, 60, 61, 62, 63
terminals, 48
territory, 58
terrorism, 69
test procedure, 56
testimony, xi, 9, 42, 44, 61
Texas, 4, 7, 59, 62, 77
threat, 81
threatened, 77
threatening, 15, 78
threats, vii, xii, 65
threshold, 81
thresholds, 81
time lags, 6
timing, 15
title, 10
total costs, 42
total energy, 43
trade, 5, 12, 26, 29, 58, 63, 69, 76
trading, 7, 14, 15, 79
transactions, 7, 26, 43, 44, 48, 50
transfer, 7
transformation, viii, 2
transition, 57
transition period, 57
transmits, ix, 18
transparency, 44, 54, 59
transport, 55
transportation, vii, viii, ix, x, xi, 1, 2, 18, 20, 22, 26, 27, 31, 32, 33, 35, 36, 37, 38, 39, 68, 75, 83
travel, 39, 61
Treasury, 33, 38, 39
trucking, 33, 39, 45, 62
trucks, 37, 48, 66, 68, 79
trust, 36, 37, 38, 39
trust fund, 36, 37, 38, 39

U

UAE, 13
uncertainty, vii, x, xii, 31, 65
uniform, 50
United Kingdom, 45, 56, 58, 62
United States, ix, x, 5, 10, 14, 15, 16, 18, 20, 21, 25, 27, 28, 29, 35, 36, 43, 45, 47, 48, 50, 59, 61, 62, 71, 76, 79, 80
United States Postal Service, 61
Utah, 76

V

vacation, 39
validity, 9
values, 22, 29
variance, 57
vehicles, 37, 63, 79
Venezuela, 11, 13
vessels, 48
violence, vii, xii, 15, 65, 69
visible, vii
volatility, vii, xii, 5, 65, 68, 74
volumetric changes, 48
voting, 50

W

war, vii, xii, 65
warrants, 78
wealth, 9, 12
well-being, viii, 2
White House, 77, 79
wholesale, 43, 48
wholesalers, 48
winter, 27, 32

writing, 38
Wyoming, 62, 76

yield, 5, 19, 20